Developed and Published
by
AIMS Education Foundation

**Research and Development**
- Betty Cordel
- Evalyn Hoover
- David Mitchell
- Myrna Mitchell
- Sheryl Mercier
- Michelle Pauls
- Dave Youngs

**Desktop Publisher**
- Leticia Rivera

**Illustrators**
- Brock Heasley
- Ben Hernandez
- Renee Mason
- Dawn McAndrews
- Sheryl Mercier
- Margo Pocock
- Dave Schlotterback
- Brenda Wood

**Developed and Published**
**by**

## AIMS Education Foundation

This book contains materials developed by the AIMS Education Foundation. **AIMS** (**A**ctivities **I**ntegrating **M**athematics and **S**cience) began in 1981 with a grant from the National Science Foundation. The non-profit AIMS Education Foundation publishes hands-on instructional materials that build conceptual understanding. The foundation also sponsors a national program of professional development through which educators may gain expertise in teaching math and science.



AIMS Education Foundation
P.O. Box 8120, Fresno, CA 93747-8120 • 888.733.2467 • aimsedu.org

ISBN 978-1-60519-065-5

Printed in the United States of America

# Earth Rocks!

## Table of Contents

# I Hear and I Forget,

# I See and I Remember,

# I Do and I Understand.

**-Chinese Proverb**

 

# Assembling Rubber Band Books

Rubber band books offer valuable content information in a kid-friendly way. Each student can be given his or her own book to keep and refer to at a later date. These books also provide a great home link, as students can take them home and share the information they are learning with their parents. To assemble a book, follow these simple instructions:

A #19 rubber band fits perfectly. If these are not available, snip the top and bottom of the center fold line of the book so that the other rubber bands can fit.

 

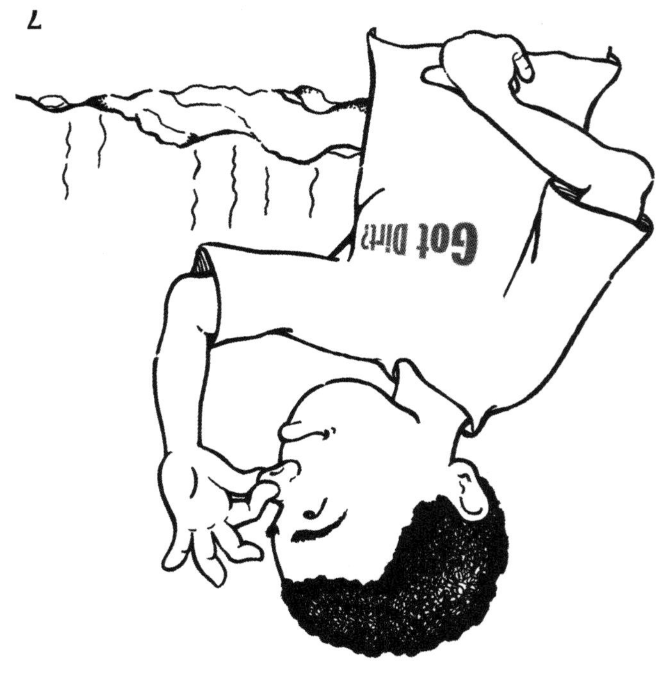


Humus is the name given to the organic materials that come from the decaying plants and animals that are in soil. Humus adds the richness to soil as well as the distinct smell some soils have.



Geologists investigate materials of the Earth's crust such as soil, sand, and rocks.

Rocks come in a large variety of shapes, textures, and colors.

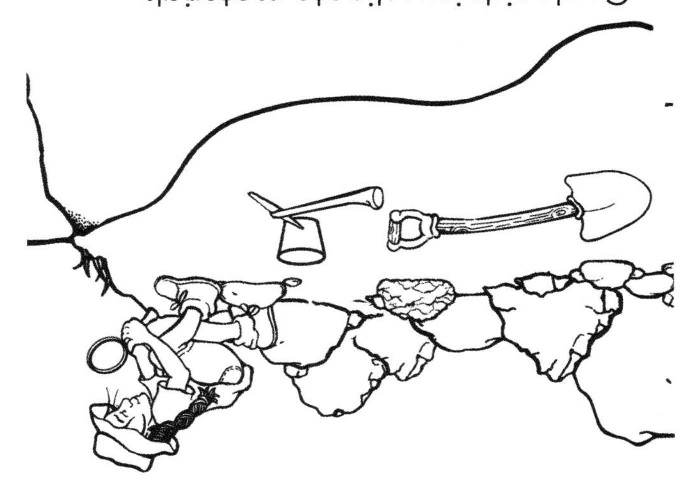

Rocks and soils play a very important part in our lives. How many ways can you list?

8

# The Amazing Geosphere

The geosphere is what scientists call the solid part of our Earth. This solid part plays a very important part of our everyday lives.

1

 

Rocks are solid objects that make up Earth's surface. Rock underlies the hills and mountains. It is under the ocean and under ice cap at the South Pole.

3

4

Layers of rocks can be seen where roads have been cut through hills. Rocks can be seen all over the Earth.

5

Rocks are slowly broken down into smaller pieces by a process called weathering. Soil is a product of this process.

Soil needs another ingredient besides rocks. It needs organic materials. Organic materials are things that are alive or once were alive.

6

1

Rocks are classified into three groups. The three groups are:

- sedimentary

- igneous

- metamorphic

2

Rocks are classified by the way they are formed. You will study the ways rocks are formed.

3

Rocks are always changing. They can change from one type to another. This constant changing is called the rock cycle.

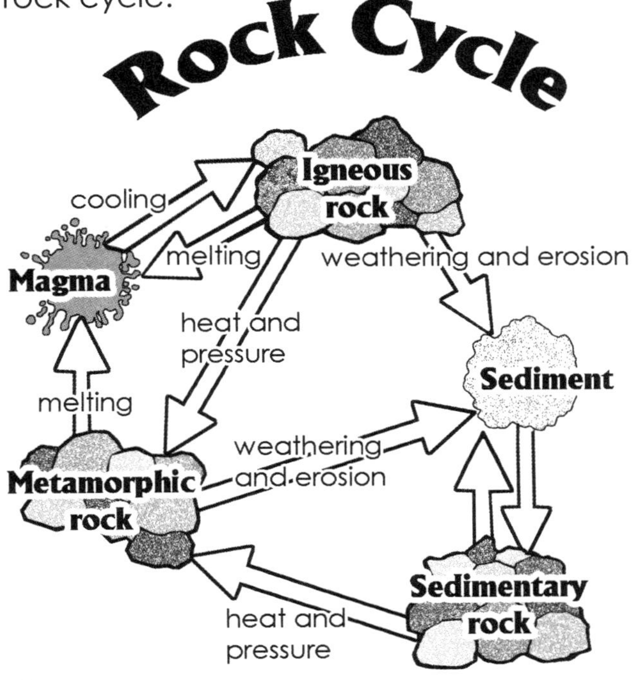


4

# Some are Sedimentary

**Topic**
Sedimentary rocks

**Key Question**
How are sedimentary rocks formed?

**Learning Goal**
Students will model how some sedimentary rocks form.

**Guiding Documents**
*Project 2061 Benchmarks*
- *Rock is composed of different combinations of minerals. Smaller rocks come from the breakage and weathering of bedrock and larger rocks. Soil is made partly from weathered rock, partly from plant remains—and also contains many living organisms.*
- *Models are often used to think about processes that happen too slowly, too quickly, or on too small a scale to observe directly, or that are too vast to be changed deliberately, or that are potentially dangerous.*

*NRC Standard*
- *Some changes in the solid earth can be described as the "rock cycle." Old rocks at the earth's surface weather, forming sediments that are buried, then compacted, heated, and often recrystallized into new rock. Eventually, those new rocks may be brought to the surface by the forces that drive plate motions, and the rock cycle continues.*

**Science**
Earth science
  sedimentary rocks

**Integrated Processes**
Observing
Comparing and contrasting
Collecting and recording data
Communicating

**Materials**
*For each group:*
  empty one-liter bottle with lid
  soil, sand, and pebbles (see *Management 1)*
  funnel (see *Management 2)*
  sedimentary rocks (see *Management 3)*
  hand lenses
  water source

*For the teacher:*
  *Sedimentary Rock Information* cards
    (see *Management 4)*

**Background Information**
Sedimentary rocks are made up of materials that were once a part of other rocks, minerals, and/or once-living things. These parts, called sediments, were deposited as layers of loose materials. Most sediments are deposited on ocean floors or at the bottom of rivers and lakes. Over time, the loose sediments are formed into rocks.

Sedimentary rocks are classified into three groups. They are grouped based on what the original sediments were. Clastic sedimentary rocks form from layers of sediments. Pressure causes the water around the sediments to be squeezed out and the sediments are cemented together. Sandstones, shales, and conglomerates form this way.

Chemical sedimentary rocks form when rock fragments dissolve in water. Over time, the water evaporates and the minerals that were in the rock crystallize into large deposits. Rock salt and gypsum form this way.

Organic sedimentary rocks form from the sedimentary remains of shells, skeletons, and other plant and animal parts. Limestones and coals form this way.

**Management**
1. You will need to gather builder's sand, potting soil, and small pebbles. These will be called *Earth materials* as well as *sediments* in this activity. Students will need to understand that these materials come from rocks that have been broken down as a result of weathering. The soil also contains organic materials.

 

2. You will need funnels that will permit the students to easily place the sand, soil, and pebbles into the bottles. Paper funnels sold to pour oil into cars are an inexpensive source for these.
3. Provide samples of conglomerates, breccias, sandstones, limestones, and shales for each group. Rock sets (item number 4108R) are available from AIMS. Hand lenses (item number 1977) are also available from AIMS.
4. The *Sedimentary Rock Information* will provide you with interesting information, some of which you may wish to tell students. You may want students to look at these rocks to see if they can find evidence of the minerals in them.

**Procedure**

*Part One*

1. Ask the *Key Question* and state the *Learning Goal.*
2. Show the students the three Earth materials. Direct them in a discussion about these materials. Make sure they understand that these materials are called *sediments.*
3. Point out to the students that these three materials came from other materials. The sand and pebbles are smaller pieces of rocks that have been broken down by processes on the Earth. The soil is a combination of sand and rocks as well as organic materials. Organic means that it was once living. Examples of organic materials found in the soil are decaying roots and leaves, as well as insects.
4. Distribute the student sheet and have the students read and follow the directions.
5. Assist students through *Part One.*
6. Point out the location for the students to place their bottles so that they can settle out over night.

*Part Two*

1. Ask the *Key Question* and state the *Learning Goal.*
2. Review the procedure from *Part One.* Allow time for the students to make observations. Make sure they notice that the sediments have formed layers—larger particles are on the bottom, smaller particles on top.
3. Distribute the rubber band book *Sedimentary Rocks.* Allow time for reading and discussion.
4. Distribute the sedimentary rock samples and hand lenses.
5. Have students relate the particles to the sediments in their bottles. The larger particles in their bottles represent the conglomerate and breccia rock samples. The smaller particles in the bottles represent the sandstone, shale, and limestone.

**Connecting Learning**

1. In your own words, describe how some sedimentary rocks form.
2. What can you infer about how the weight of sediments affects the formation of sedimentary rocks?
3. What role does water play in the formation of sedimentary rocks?
4. How are the sandstone, shale, and limestone the same? How are they different?
5. How are the conglomerate and breccia the same? How are they different?
6. What are you wondering now?

 

# Sedimentary Rock Information

## Sandstone

**Type of rock**
Sedimentary

**Texture**
Fine, sand-sized grains

**Minerals common to it**
Quartz and feldspar

**Colors**
White, orange, pink, and red

**Where it forms**
Sandstones form in regions where quartz sand particles are deposited and buried under layer after layer. This deposition can take place along the coast, in a river, or in a desert.

**Fascinating Factoids**
Sandstones are often used by geologists to tell about what past environments of the Earth were like. Deep layers of sandstone tell us that this region was once a desert. Horizontal layers show how the sand accumulated in calm water. Aquifers can often be found in sandstone rock formations. Aquifers in sandstones filter out pollutants. Sandstones are often used as building materials.

## Breccia

**Type of rock**
Sedimentary

**Texture**
Large pieces of rocks and sediments that have sharp edges

**Minerals common to it**
Almost any mixture of any minerals

**Colors**
A wide variety of colors

**Where it forms**
Breccias form when rock fragments are cemented together with silicas, calcite, and iron oxides. The rock particles in breccias are sharper than those in conglomerates because water, wind, or glaciers have not transported them as long. Breccias often form nearest the land where a river empties into an ocean or large body of water.

**Fascinating Factoids**
The Greeks used breccias as a decorative building material. Breccias today are often used to make jewelry.

# Sedimentary Rock Information

## Fossiliferous Limestone

**Type of rock**
Sedimentary

**Minerals common to it**
Calcite, dolomite, and aragonite

**Texture**
Fossils of all shapes and sizes with fine-grained calcite in between

**Colors**
White, gray, black, and pink

**Where it forms**
Limestone forms in a shallow sea when layers of shells and skeletons of small marine animals are buried and compressed. Over time, the layers turn to rock.

**Fascinating Factoids**
Limestone often contains fossils. Chalk is a soft, fine-grained fossiliferous limestone that is composed of the remains of tiny marine shells called foraminifera. The main use of limestone is to make cement. Limestone has many other uses. It is used to make lime and in the manufacture of paper. It is also used in insecticides, linoleum, and fiberglass. It is used in the backing for some carpets.

## Shale

**Type of rock**
Sedimentary

**Minerals common to it**
Quartz

**Texture**
Fine-grained

**Colors**
Gray, black, green, and red

**Where it forms**
Shale forms when tiny clay particles settle on the bottom of bodies of water. The clay particles are pressed together by the pressure from the weight of the other particles above as well as the weight of the water.

**Fascinating Factoids**
Shale often contains fossils. Some shale contains oil. Shale can be finely ground and used as filler in paints, plastics, asphalt compounds, roofing cement, and some linoleum.

# Sedimentary Rock Information

## Conglomerate

**Type of rock**
Sedimentary

**Minerals common to it**
Almost any mixture of any minerals

**Texture**
Large pieces of rocks and sediments that have rounded edges

**Colors**
A wide variety of colors

**Where it forms**
Conglomerates form when rock fragments are cemented together with silicas, calcite, and iron oxides. The rock particles in conglomerates are rounded. This is a result of them being transported by water, wind, or glaciers. Conglomerates form when a river empties into an ocean or large body of water.

**Fascinating Factoids**
Conglomerates are used in building roads as well as the roadbed for railroad tracks. Some conglomerates are used for decorative purposes. Conglomerates often look like broken pieces of concrete.

 

# Some are Sedimentary

## Key Question

How are sedimentary rocks formed?

## Learning Goal

**Students will:**

model how some sedimentary rocks form.

## Modeling Sedimentary Rocks

**Part One**

1. Add soil, sand, and pebbles to the bottle.
2. Add water and shake.
3. Wait five minutes.
4. Observe, sketch, and describe what you see.
5. Let the bottle set overnight.

**Part Two**

1. Observe, sketch, and describe what you see.

2. In your Science Journal, describe how this shows one way sedimentary rocks form.

 

# All About Sedimentary Rocks

**Sedimentary rocks** are made of sediments.

Sediments can be lots of things. They can be pieces of other rocks. They can be minerals. They can be parts of once-living things.

1

Sediments are moved by water and wind. They pile up in layers.

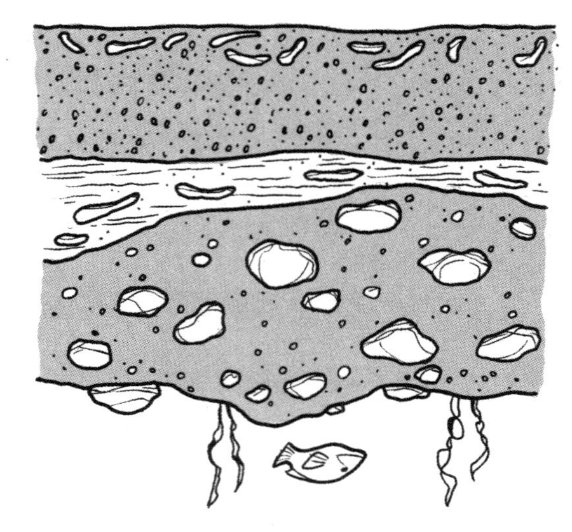

Sometimes these layers are on land. Often they are at the bottom of an ocean, river, or lake. Over time, the layers of sediment get pressed together and become rock.

2

There are different kinds of sedimentary rock. Some are made up of different size sediments.

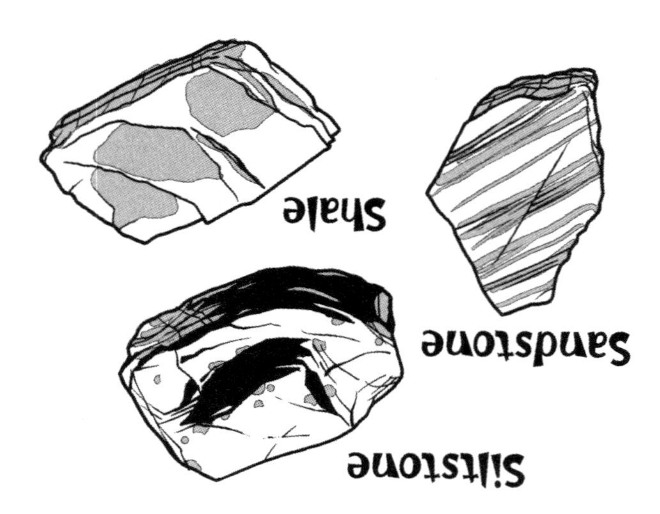


Sandstone, shale, and siltstone are this kind of rock.

3

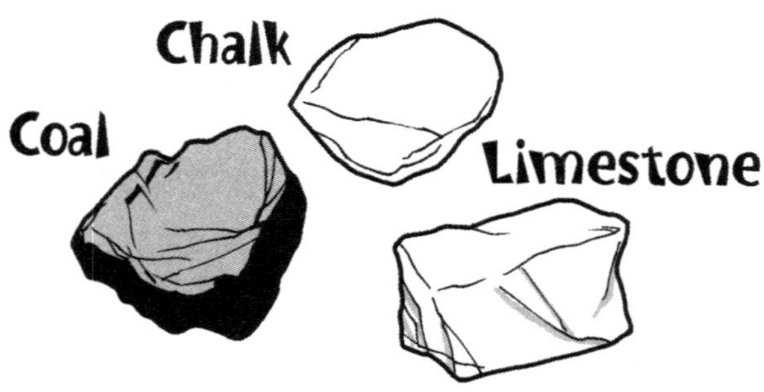


Others are made of once-living things. Coal, chalk, and most limestone are this kind of rock.

Some are made from minerals left behind when water evaporates. Halite and gypsum are this kind of sedimentary rock.

4

# Sedimentary Rock Sources

## Connecting Learning

1. In your own words, describe how some sedimentary rocks form.
2. What can you infer about how the weight of sediments affects the formation of sedimentary rocks?
3. What role does water play in the formation of sedimentary rocks?
4. How are the sandstone, shale, and limestone the same? How are they different?
5. How are the conglomerate and breccia the same? How are they different?
6. What are you wondering now?

**Topic**
Igneous rocks

**Key Question**
How does the cooling rate of a solution affect the size of the crystals that will form?

**Learning Goals**
Students will:
- model how some igneous rock form; and
- observe, compare, and contrast samples of igneous rocks.

**Guiding Documents**
*Project 2061 Benchmarks*
- *Rock is composed of different combinations of minerals. Smaller rocks come from the breakage and weathering of bedrock and larger rocks. Soil is made partly from weathered rock, partly from plant remains—and also contains many living organisms.*
- *Models are often used to think about processes that happen too slowly, too quickly, or on too small a scale to observe directly, or that are too vast to be changed deliberately, or that are potentially dangerous.*

*NRC Standard*
- *Some changes in the solid earth can be described as the "rock cycle." Old rocks at the earth's surface weather, forming sediments that are buried, then compacted, heated, and often recrystallized into new rock. Eventually, those new rocks may be brought to the surface by the forces that drive plate motions, and the rock cycle continues.*

**Science**
Earth science
igneous rock

**Integrated Processes**
Observing
Comparing and contrasting
Relating
Communicating
Collecting and recording data
Interpreting data
Inferring

**Materials**
*For each student group:*
3 baby food jars
3 chenille stems
3 index cards
150 mL of supersaturated Epsom salt solution (see *Management 1)*
hand lens
igneous rock samples (see *Management 4)*
*Igneous Rock Study Print*

*For each student:*
*Cool Crystals* rubber band book
student pages

*For the class:*
refrigerator (see *Management 2)*
cookie sheet

*For the teacher:*
*Igneous Rock Information* cards

**Background Information**
The word *igneous* means born from fire. Igneous rock is formed from magma that has hardened and crystallized. Magma is molten rock that also contains early-formed crystals and dissolved gases. Lava is magma that reaches the surface of the Earth, usually through erupting volcanoes.

Geologists classify igneous rocks into two main types—intrusive and extrusive. Intrusive igneous rocks form underground where they cool slowly. This often gives them a coarse texture with large mineral crystals. Granite and gabbro are examples of intrusive igneous rocks.

Extrusive igneous rocks form when magma reaches the surface of the Earth or the ocean floor, cooling rapidly. This causes them to have very small crystals. Basalt, obsidian, and pumice are examples of extrusive igneous rocks.

In terms of mineral composition, intrusive and extrusive igneous rocks can be the same. For example, basalt, which makes up much of the ocean floor, is identical in composition to gabbro, which is formed deep in the Earth's crust. The former is extrusive and the latter is intrusive, so they have very different appearances despite their identical compositions.

 

**Management**

1. Create a supersaturated solution of Epsom salts and water. Heat a pan of water until it nearly boils. Stir in Epsom salts until no more will dissolve. You will need to prepare enough of the solution so that each group will have 150 mL of solution.
2. Make arrangements to place one jar from each group into a refrigerator. A cookie sheet will provide easy transport for the baby food jars to the refrigerator.
3. Begin this activity on a Monday so that five days of observations can be made.
4. You need samples of igneous rocks that include granite, gabbro, and basalt. If possible, each group should have its own set of samples. A rock set that includes these samples is available from AIMS (item number 4108R). Hand lenses (item number 1977) are also available from AIMS.

**Procedure**

*Day One*

1. Ask the *Key Question* and state the *Learning Goals*.
2. Distribute the *Cool Crystals* rubber band book to each student and read through the information as a class.
3. Point out to the students that they cannot directly observe how magma cools to form crystals, but they can explore how cooling affects the formation of crystals in a solution.
4. Divide students into groups and distribute the Epsom salt solution, baby food jars, index cards, and chenille stems to each group. Have the group members write their names on all of the index cards.
5. Direct the students to pour 50 mL of the solution into each of the baby food jars. No precise measurements are necessary; students can merely divide the original amount equally into the three jars.
6. Show them how to create a spiral with the chenille stem by wrapping it around a pencil. Tell students to place one spiral into each of the three baby food jars with the solution.
7. Inform the students that they will be placing their three baby food jars with the solution in three different temperature locations. They will test them at room temperature, in a refrigerator, and in a location of their choosing.
8. Give each student a copy of the first student page. Have groups decide on locations for their third jars and record these on the page.
9. Point out that they will need to make observations of each of the containers each day for five days using drawings as well as written observations. Distribute hand lenses and instruct them to record their initial observations of each jar.
10. Have each group place one of its index cards on the cookie sheet with one of its jars on top of the card. Take the cookie sheet and place it in the refrigerator.
11. Instruct groups to place another of their jars on top of an index card in an area in the classroom that will not be disturbed.
12. Allow groups to place their third jars in locations of their choosing on top of the third index card.

*Days Two-Four*

1. Remove the cookie sheet from the refrigerator and have groups collect their jars.
2. Allow time for groups to make and record observations on their student pages. Encourage them to use the hand lenses for more detailed observations.
3. Repeat the observations with the two remaining jars.
4. Return all of the jars to their initial locations.

*Day Five*

1. Have groups make one final observation of each of their jars.
2. Distribute the igneous rock samples to each group along with the second student page.
3. Allow time for students to make observations of each rock sample and record those observations on the student page.
4. Distribute the *Igneous Rock Study Print* to each group. Guide the students in placing the granite, gabbro, and basalt in the correct spaces. The granite and gabbro may be placed in the first or second circle. The basalt needs to be placed in the third circle.
5. Distribute the final student page and have students respond to the questions. Discuss what they learned about igneous rocks from this activity.

**Connecting Learning**

1. In your own words, describe how some igneous rocks form.
2. What do you notice about the crystals from the three different locations?
3. Why do you think they are different? Which one cooled more slowly?
4. How might the crystals or grains in an igneous rock differ because of how quickly the molten rock cooled?
5. Look at the igneous rock samples you have. Which do you think were cooled slowly? Why?
6. Why did we need to model how igneous rocks formed? What did the model show us? What didn't the model show us?
7. What are you wondering now?

 

# Igneous Rock Information

## Granite

**Type of rock**
Igneous

**Texture**
Coarse-grained

**Minerals common to it**
Quartz, mica, feldspar

**Colors**
Light gray, salmon pink, white, dark minerals

**Where it forms**
Granite is extrusive igneous rock, forming under the surface of the Earth where it cools slowly allowing larger crystals to form. Some crystals can be a centimeter or more in length.

**Fascinating Factoids**
Granite is the most common rock on continental surfaces. It is less dense than basalt (which is the most common rock on Earth and is found on the ocean floors—the oceanic crust), so granite tends to "float" higher on the Earth's mantle. Granite is resistant to weathering and erosion so is often found as the core of many mountains.

**Uses**
Facings on buildings, monuments, headstones, kitchen countertops, statues

## Basalt

**Type of Rock**
Igneous

**Texture**
Coarse-grained

**Minerals common to it**
Feldspar, pyroxene, olivine

**Colors**
Dark grays and black

**Where it forms**
Basalt is an extrusive igneous rock, usually forming in the ocean waters where it cools rapidly thus limiting crystal growth.

**Fascinating Factoids**
Basalt is the most common rock found on Earth's surface. It is found on the ocean floors—the oceanic crust). Basaltic rocks contain a high percentage of iron and magnesium, making them dark and heavy. The dark areas on the moon are basaltic in nature.

**Uses**
Construction, crushed basalt is used as ballast for railroad tracks

 

# Igneous Rock Information

## Gabbro

**Type of Rock**
Igneous

**Minerals common to it**
Pyroxene, olivine, feldspar

**Texture**
Coarse-grained

**Colors**
Dark green to black

**Where it forms**
Gabbro is an intrusive igneous rock that has cooled slowly, giving it coarse-grained crystals. Gabbro has the same composition as basalt. Gabbro forms underneath the basalt.

**Fascinating Factoids**
Gabbro and basalt are family members. One can tell the difference in the two because gabbro has a coarse grain due to its slower cooling time. Gabbro is the "deep version" of basalt. It does not come in contact with the ocean waters like basalt. (The ocean waters cool basalt more quickly.) Gabbro stands up well to weathering and erosion. It is often called "black granite."

**Uses**
Headstones, kitchen countertops, floor tile

## Andesite Porphyry

**Type of Rock**
Igneous

**Minerals common to it**
Pyroxene, feldspar, hornblende, biotite

**Texture**
Fine-grained

**Colors**
Medium to dark gray

**Where it forms**
Andesite Porphyry is an extrusive igneous rock. (Porphyry is used to describe rocks with more than one size of mineral fragments.) Andesite is fine-grained. Yet within it are visible crystals that had a different cooling rate.

**Fascinating Factoids**
Andesite gets it name from the Andes mountains where it is common. Explosive volcanic eruptions often produce andesite, such was the experience with the huge eruption of Krakatoa in 1883. Andesite is slip resistant.

**Uses**
Tiles or rocks around water gardens.

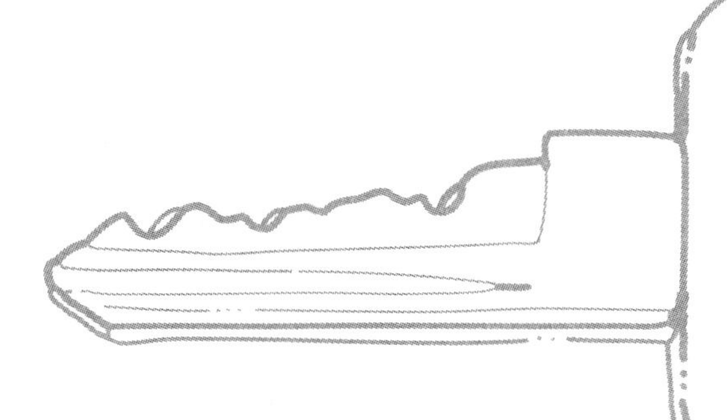

## Key Question

How does the cooling rate of a solution affect the size of the crystals that will form?

## Learning Goals

**Students will:**

- model how some igneous rock form; and
- observe, compare, and contrast samples of igneous rocks.

The word *igneous* means born from fire. Igneous rock is formed from magma that has hardened and crystallized.





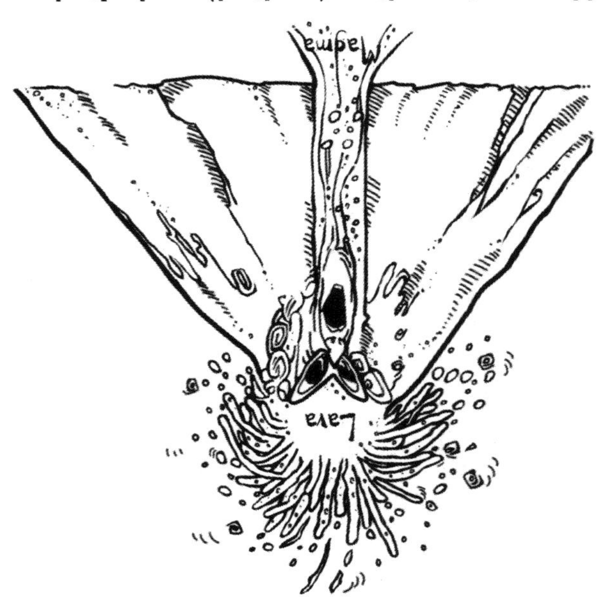


Magma is molten (melted) rock. It also contains gases and crystals. Lava is magma that reaches the surface of the Earth. This usually happens through erupting volcanoes.



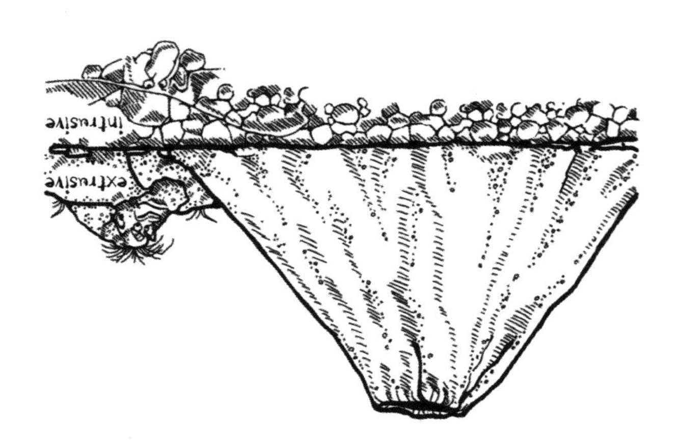


There are two kinds of igneous rocks—intrusive and extrusive. Intrusive igneous rocks form underground. They cool slowly. This often gives them a coarse texture with large mineral crystals. Granite and gabbro are examples of intrusive igneous rocks.

Extrusive igneous rocks form when magma reaches the surface of the Earth or the ocean floor. Extrusive igneous rocks cool rapidly. This causes them to have very small crystals. Basalt, obsidian, and pumice are extrusive igneous rocks. Basalt covers much of the ocean floor.

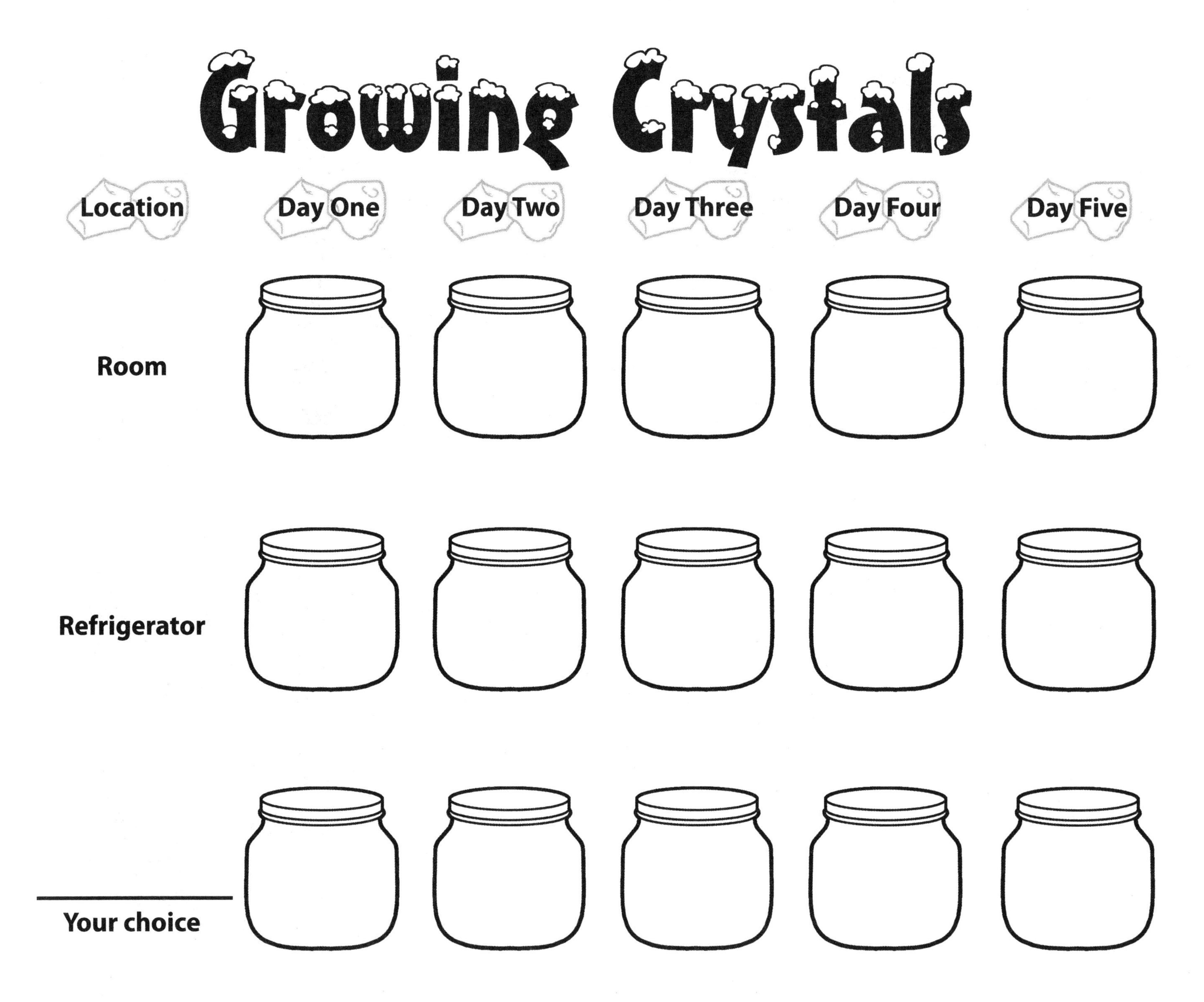
Growing Crystals
Location
Day One
Day Two
Day Three
Day Four
Day Five
Room
Refrigerator
Your choice

 

# Growing Crystals

Record information about each of your rock samples. Include a sketch of each; its name; observations of color, texture, etc.; and how it is alike and different than the other rock samples.

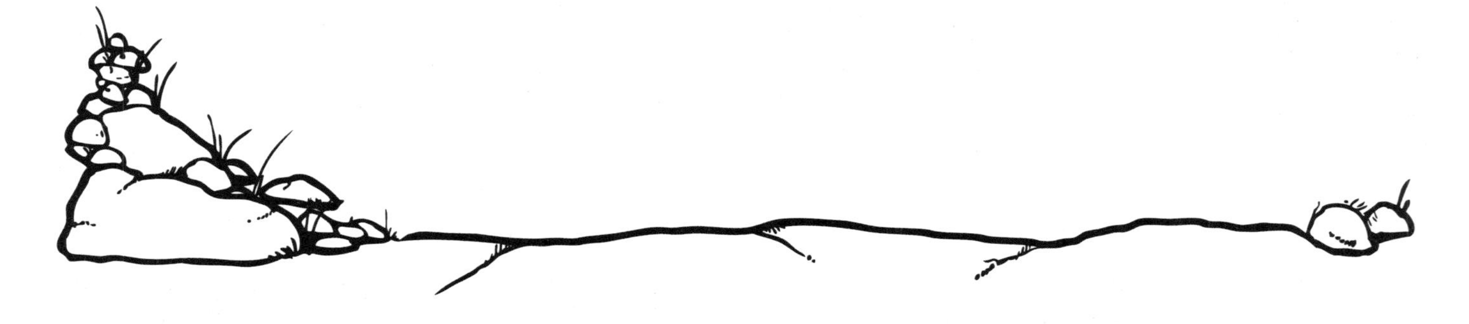

 

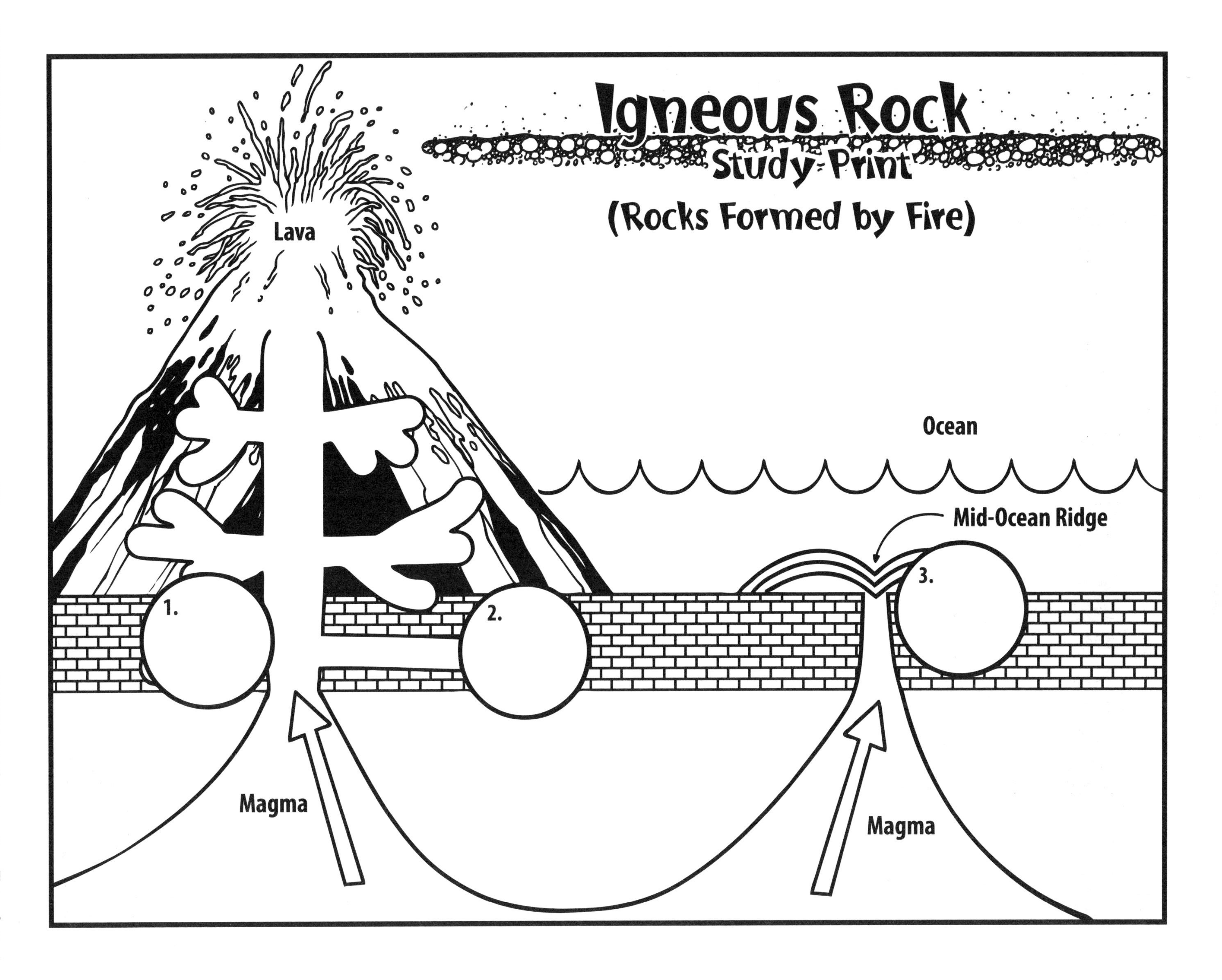

Igneous Rock
Study Print
(Rocks Formed by Fire)
Lava
Ocean
Mid-Ocean Ridge
1.
2.
3.
Magma
Magma


 

# Growing Crystals

Answer these questions after completing the activity.

1. What does the word *igneous* mean?

2. Do you think *igneous* is a good way to describe these kinds of rocks? Explain.

3. Explain the difference in how intrusive and extrusive igneous rocks form.

4. How can the size of the crystals in an igneous rock help you identify how it was formed?

5. What is the difference between magma and lava?

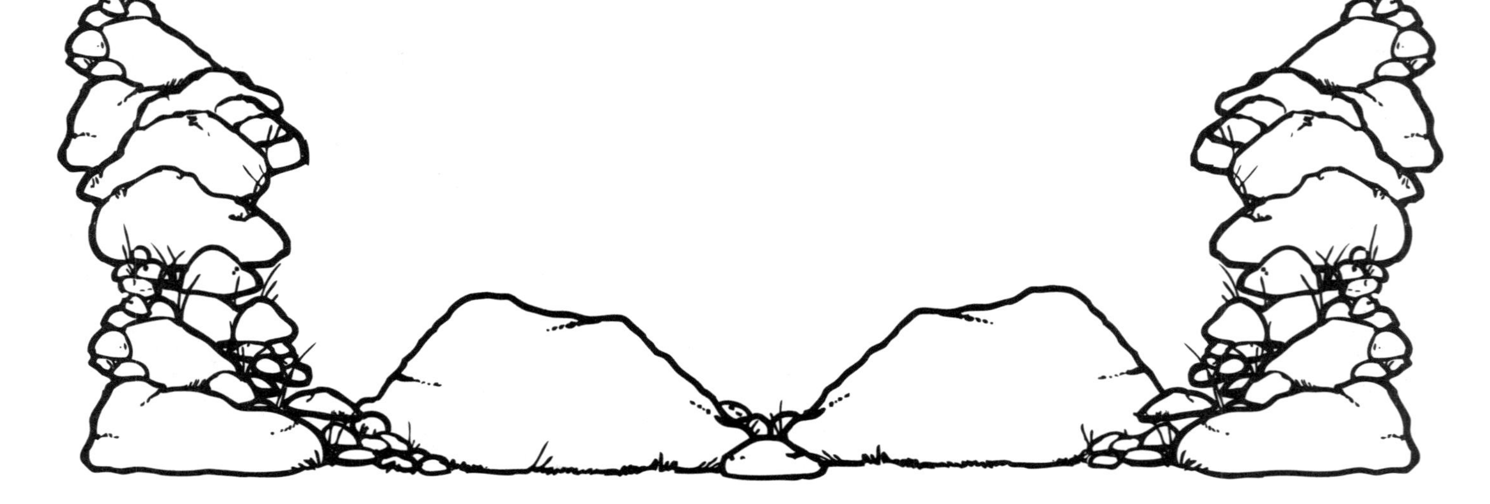

 

# Growing Crystals

## Connecting Learning

1. Describe how some igneous rocks form.

2. What do you notice about the crystals from the three different locations?

3. Why do you think they are different? Which one cooled more slowly?

4. How might the crystals or grains in an igneous rock differ because of how quickly the molten rock cooled?

 

# Growing Crystals

CONNECTING LEARNING

## Connecting Learning

5. Look at the igneous rock samples you have. Which do you think were cooled slowly? Why?

6. Why did we need to model how igneous rocks formed? What did the model show us? What didn't the model show us?

7. What are you wondering now?

**Topic**
Metamorphic rocks

**Key Question**
How can a model help us understand more about how metamorphic rocks form?

**Learning Goals**
Students will:
- model how some metamorphic rocks form, and
- make observations of metamorphic rocks.

**Guiding Documents**
*Project 2061 Benchmarks*
- *Rock is composed of different combinations of minerals. Smaller rocks come from the breakage and weathering of bedrock and larger rocks. Soil is made partly from weathered rock, partly from plant remains—and also contains many living organisms.*
- *Models are often used to think about processes that happen too slowly, too quickly, or on too small a scale to observe directly, or that are too vast to be changed deliberately, or that are potentially dangerous.*

*NRC Standard*
- *Some changes in the solid earth can be described as the "rock cycle." Old rocks at the earth's surface weather, forming sediments that are buried, then compacted, heated, and often recrystallized into new rock. Eventually, those new rocks may be brought to the surface by the forces that drive plate motions, and the rock cycle continues.*

**Science**
Earth science
metamorphic rocks

**Integrated Processes**
Observing
Comparing and contrasting
Relating
Communicating
Collecting and recording data
Inferring

**Materials**
*For each student:*
1 piece of cookie bar mixture (see *Management 1)*
1 piece of heavy duty aluminum foil, 10 cm by 10 cm
2 pieces of cardboard,10 cm by 10 cm
metamorphic rocks (see *Management 3)*
*All About Metamorphic Rocks* mini-book

*For the class:*
1 clothes iron

*For the teacher:*
*Metamorphic Rock Information* cards
(see *Management 4)*

**Background Information**
Metamorphic rocks are formed when rocks are changed due to heat or pressure. These changed rocks are usually much harder than the original rocks. Metamorphic rocks may be foliated or banded. They may look like they have layers. The "stripes" form because of the different minerals in the rock. These minerals have been pressed into bands by heat and pressure.

The mineral structure and form of an igneous rock or sedimentary rock can change, or metamorphose, when the rock is heated by contact with magma or pressure underground. This contact does not cause the rock to melt. The heat and pressure can cause chemical changes in the rock as well as change the mineral structures in the rock. Granite, an igneous rock, becomes gneiss as a result of this heat and pressure. The minerals are the same in both rocks, but the rocks look very different.

Some metamorphic rocks look like they have layers similar to those you can see in sedimentary rocks. The minerals in the rock cause these "layers" to line up in the same direction when they are put under great pressure. When shale, a sedimentary rock, is put under great pressure, it becomes the metamorphic rock slate.

**Management**

1. Use the following ingredients and recipe to make the cookie bar.

   Ingredients:
   1/4 cup of butter or margarine
   5 cups of crispy rice cereal
   4 cups of miniature marshmallows
   1 cup of raisins
   1/2 cup of toffee chips
   1/2 cup of candy-coated chocolate candies

   Melt the butter or margarine in a saucepan. Add the marshmallows and cook over low heat, stirring until the marshmallows are melted and the mixture is syrupy. Remove from the heat and stir in the cereal. Stir in the toffee chips. Firmly press the warm mixture into a greased 9- by 12-inch pan. Evenly distribute the chocolate candies and raisins on the surface of the mixture and gently press them in. When the mixture cools, cut it into enough pieces so that each student will have a single piece and enough "control" pieces are left for each group to compare their changed pieces with the controls.
2. You will need to select a place to let the students place the cookie bars so that you can apply the heat with the iron. It is suggested that you complete the heating of the cookie bars when the students are NOT in the room. Press down on each aluminum foil packet with the clothing iron set on the highest setting. Heat each package for about 20 seconds. The aluminum foil can get very hot and cause burns. Allow the cookie bars to cool completely before the students make observations.
3. You will need a collection of metamorphic rocks. You need to include gneiss, slate, and schist. These clearly show some of the classic properties of metamorphic rocks. A set of rocks (item number 4108R) is available from AIMS.
4. The *Metamorphic Rock Information* will provide you with interesting information, some of which you may wish to tell students. You may want students to look at these rocks to see if they can find evidence of the minerals in them.

**Procedure**

*Part One*

1. Ask the *Key Question* and state the *Learning Goals*.
2. Distribute the mini-book to students. Have them fold it and read it. Discuss the information and the *Background Information* on how metamorphic rocks form.
3. Distribute a piece of the cookie bar to each student. Tell the students to record observations using both drawings and written descriptions. Make certain that they notice the different "sedements".
4. Show the students how to carefully wrap the cookie bar with the aluminum foil. Direct them to sandwich the cookie bar between the two pieces of cardboard. Show them how to place the sandwiched cookie bar on the tabletop and press down with their hands.
5. Tell them to carefully unwrap the cookie bar and record observations.
6. Guide them in rewrapping the cookie bar in the aluminum foil and placing the cookie bar back between the two pieces of cardboard.
7. Demonstrate how to stand on the cookie bar. Direct the students to carefully unwrap the cookie bar and record observations.
8. Point out to the students that these two parts of the activity used pressure. Ask them what other force is a part of the formation of metamorphic rock. [heat]
9. Show the students the iron. Tell them you will be applying heat to each of their samples. Tell students that they will examine the cookie samples after they have cooled completely.

*Part Two*

1. Direct the students to carefully unwrap the cookie bars and make observations.
2. Distribute an unchanged cookie bar to each group. Tell the students to make observations about the differences they observe in the two cookie bars.
3. Distribute the metamorphic rocks and tell the students to make observations. [The slate undergoes the least amount of change, the gneiss the most. The schist falls between the two.]

**Connecting Learning**

1. In your own words, describe how some metamorphic rocks form.
2. What metamorphic process did pressing and standing simulate in this activity? [pressure]
3. What metamorphic process did the iron simulate? [heat]
4. What did the different ingredients in the cookie bar represent?
5. Look at the metamorphic rock samples you have. What are some of the characteristics you can see in the samples?
6. Why did we need to model how metamorphic rocks formed? What did this model show us? What didn't this model show us?
7. What are you wondering now?

# Metamorphic Rock Information

## Slate

**Type of rock**
Metamorphic

**Minerals common to it**
Mica, quartz, and sometimes pyrite

**Texture**
Very fine-grained crystal arranged in flat sheets

**Colors**
Red, green, gray, and black

**Where it forms**
Slate forms when shale is changed by heat and pressure. Slate is an example of a metamorphic rock in its beginning stages of change. When slate is placed under further heat and pressure, it changes into schist and finally into gneiss.

**Fascinating Factoids**
Slate breaks into smooth flat sheets. It is very common in the mountains of North America. Slate is often used as a building material for walks as well as for roofs. It is used in classrooms as blackboards. Good quality pool tables often will have a slate surface under the felt playing surface.

## Schist

**Type of rock**
Metamorphic

**Minerals common to it**
Biotite and muscovite micas, quartz, and feldspar

**Texture**
Medium-to coarse-grained. The grains are arranged in sheets.

**Colors**
Sparkly gray to black

**Where it forms**
Schist forms in regions where rocks are under a great deal of both heat and pressure. Shale is often the beginning material for schist. As shale is placed under heat and pressure, it begins to change (metamorphose) into slate. If the slate is placed under even more pressure, it becomes schist.

**Fascinating Factoids**
Schist is foliated. That means that the individual mineral grains split off easily in to flakes or slabs. The word schist comes from Greek, and it means to split. Schist is often named for a unique mineral in it. Garnet schist contains the semi-precious gemstone garnet.

# Metamorphic Rock Information

## Gneiss

**Type of rock**
Metamorphic

**Texture**
Large grains in banded layers

**Minerals common to it**
Quartz, feldspar, biotite, and hornblende

**Colors**
Black, white, pink, and green

### Where it forms

Gneiss forms deep in the Earth's crust. It undergoes the greatest change as a result of the high heat and pressure. Granite is often the parent material for gneiss. The minerals in the granite all have different points at which they melt and recrystalize. This results in the formation of bands in gneiss.

### Fascinating Factoids

The Acasta gneiss is the oldest known crustal rock in the world. It is found in the Northwest Territories of Canada. Gneiss is very hard metamorphic rock. It is used as building materials, as well as gravel for both roads and railroad beds.

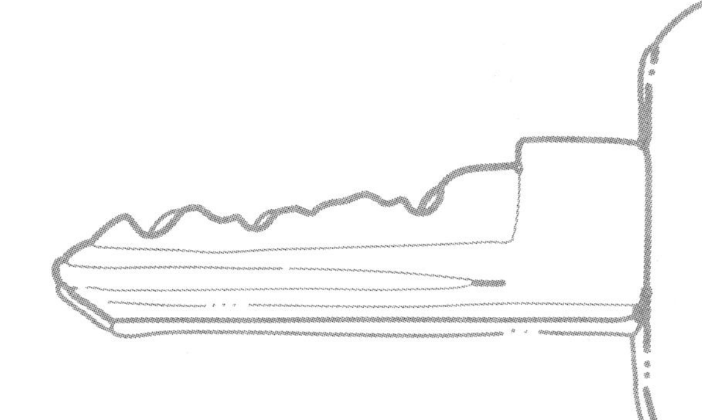

### Key Question

How can a model help us understand more about how metamorphic rocks form?

## Learning Goals

**Students will:**

- model how some metamorphic rocks form, and
- make observations of metamorphic rocks.

**Metamorphic rocks** are formed when rocks are changed by heat or pressure. This happens underground. The new rocks are usually much harder.

1

Metamorphic rocks may look layered. These "stripes" are the different minerals in the rock. The heat and pressure help them form into bands.

2

Granite is an igneous rock. It becomes gneiss with heat and pressure. The minerals in both rocks are the same. But the rocks look very different.

3

Shale is a sedimentary rock. It becomes slate with great pressure. The minerals in both rocks are the same. But the rocks look very different.

4

1. **Original cookie bar**
   Keep a record of observations.

**Write**

Draw

2. **Cookie bar changed by pressure from our hands**
   Keep a record of observations.

**Write**

**3. Cookie bar changed by pressure from standing on it**
Keep a record of observations.

**Write**

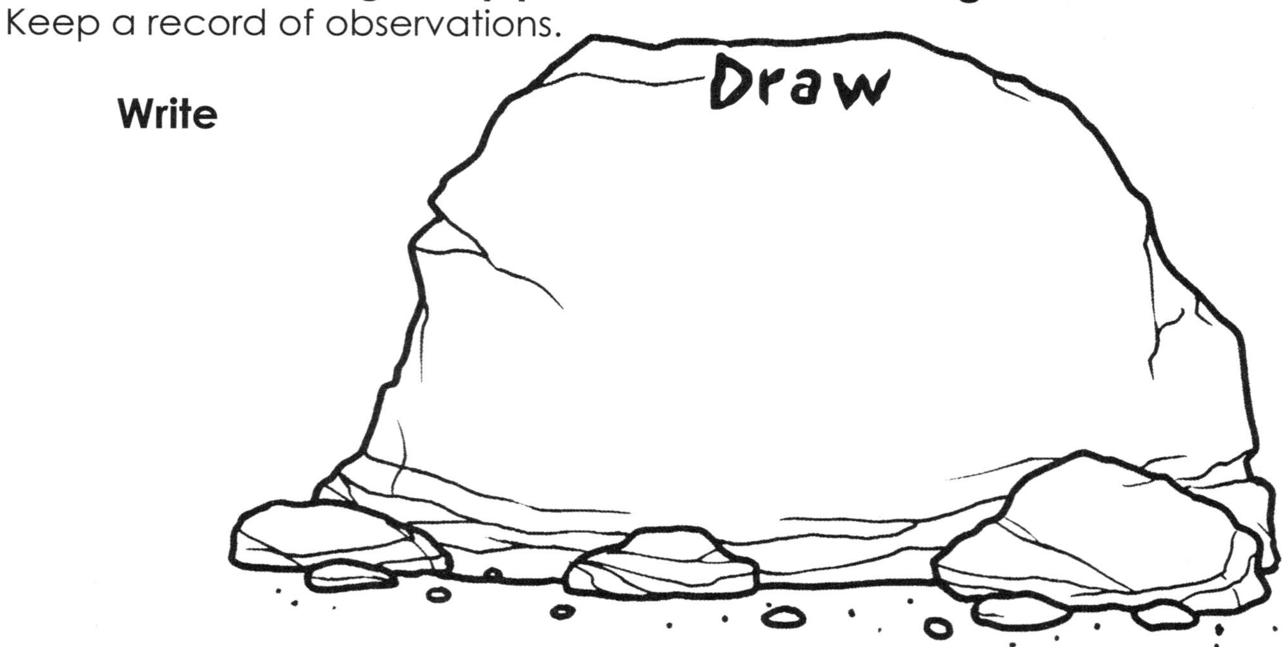


**4. Cookie bar changed by heat**
Keep a record of observations.

**Write**

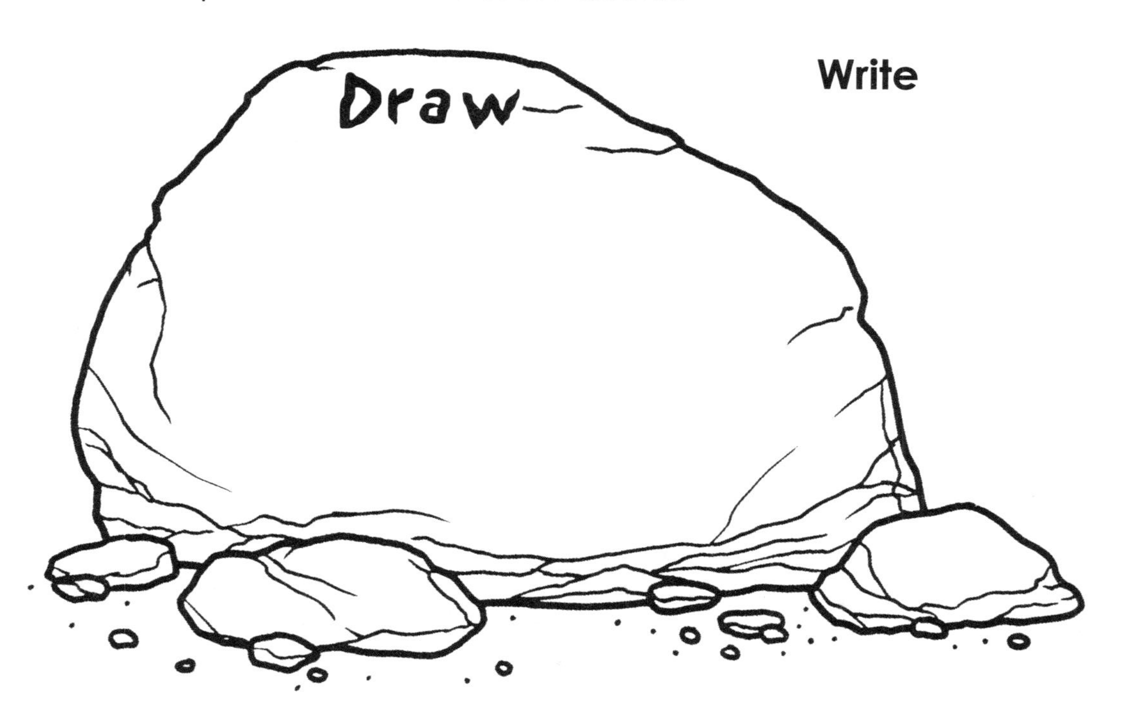

## Connecting Learning

1. In your own words, describe how some metamorphic rocks form.
2. What metamorphic process did pressing and standing simulate in this activity?
3. What metamorphic process did the iron simulate?
4. What did the different ingredients in the cookie bar represent?
5. Look at the metamorphic rock samples you have. What are some of the characteristics you can see in the samples?

CONNECTING LEARNING

## Connecting Learning

6. Why did we need to model how metamorphic rocks formed? What did this model show us? What didn't this model show us?

7. What are you wondering now?

 

# Rock Report

**Topic**
Rocks

**Key Question**
What interesting information can you find out about your rock?

**Learning Goal**
Students will do a research project on a rock of their choice.

**Guiding Documents**
*Project 2061 Benchmarks*
- *Rock is composed of different combinations of minerals. Smaller rocks come from the breakage and weathering of bedrock and larger rocks. Soil is made partly from weathered rock, partly from plant remains—and also contains many living organisms.*
- *Keep a notebook that describes observations made, carefully distinguishes actual observations from ideas and speculations about what was observed, and is understandable weeks or months later.*

*NRC Standard*
- *Earth materials are solid rocks and soils, water, and the gases of the atmosphere. The varied materials have different physical and chemical properties, which make them useful in different ways, for example, as building materials, as sources of fuel, or for growing the plants we use as food. Earth materials provide many of the resources that humans use.*

**Science**
Earth science
  rocks

**Integrated Processes**
Observing
Comparing and contrasting
Communicating

**Materials**
AIMS Rock Kit
Hand lenses
Research materials

**Background Information**
Rocks are consolidated mixtures of minerals in which the properties of the minerals are retained. Most rocks are composed of more than one mineral.

**Management**
1. Students can work alone or in small groups to conduct and report on their research.
2. You may want to assign areas to students so they can display their findings and pictures or examples of the uses of the rocks they researched.
3. There are 12 different rocks in the AIMS Rock Kit. You can allow students to select a rock of their choice or put the names of the rocks into a bag and have them draw out the rock they need to research.
4. Rock kits (item number 4108R) and hand lenses (item number 1977) are available from AIMS.

**Procedure**
1. Have students select a rock for the purpose of conducting research.
2. Discuss with the students different resources that are available.
3. Distribute the reporting page and stress that this is to serve as a guide for some of the information that they may find. Encourage them to "go beyond the handout" and provide a "better" report.
4. Urge students to be as creative as possible in their reporting and display of their findings.

**Connecting Learning**
1. What interesting things did you find out about your rock?
2. How did you describe the grain size?
3. Where was your sample from? Where else can this type of rock be found?
4. Was your rock sedimentary, igneous, or metamorphic? What does that mean? Describe the process your rock went through to become what you examined today.
5. What are some of the uses of your rock?
6. What are you wondering now?

# Rock Report

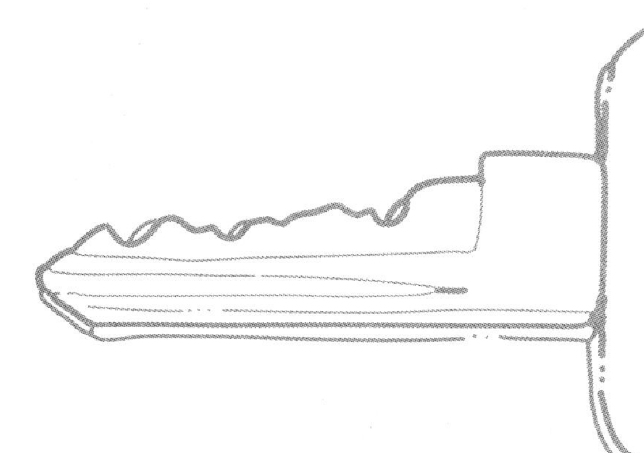

**Key Question**

What interesting information can you find out about your rock?

## Learning Goal

do a research project on a rock of their choice.

Rock Report
Name of the rock
Photograph or drawing of the actual rock
Research a rock. Record notes below.
Rock type
Grain size
Color or colors
How formed
Minerals
Uses
Other
Use your notes to write an informational report.

 

# Rock Report

CONNECTING LEARNING

## Connecting Learning

1. What interesting things did you find out about your rock?
2. How did you describe the grain size?
3. Where was your sample from? Where else can this type of rock be found?
4. Was your rock sedimentary, igneous, or metamorphic? What does that mean? Describe the process your rock went through to become what you examined today.
5. What are some of the uses of your rock?
6. What are you wondering now?

When you go outdoors for recess, you walk on the part of the Earth's surface called the crust. The crust is made up of rocks. You may say that you are walking on soil, but did you know that underneath the soil is a thick layer of rock?

1

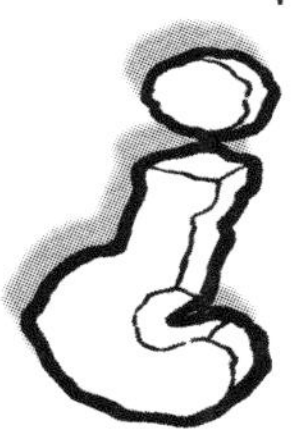

What are rocks anyway? Rocks are solid Earth materials. They are made up of a mixture of minerals.

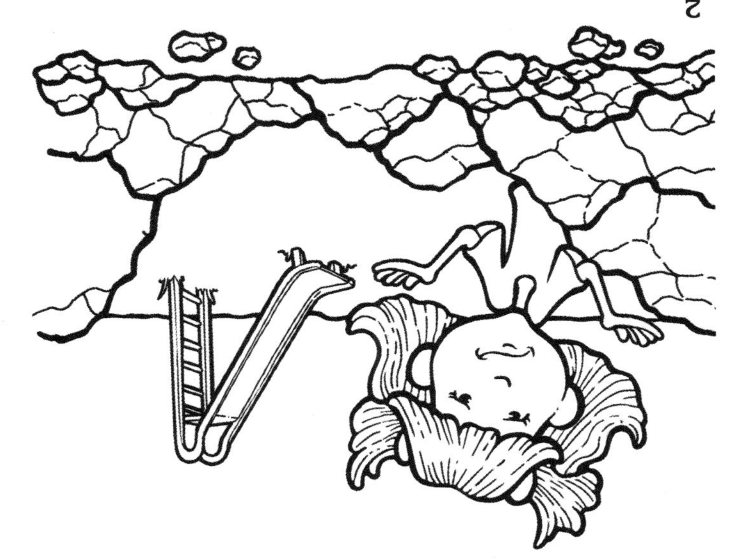

2

So, what are minerals? Minerals are solid Earth materials, too. You know many kinds of minerals. Gold is a mineral. Salt is a mineral. It is called halite. A diamond is a mineral.

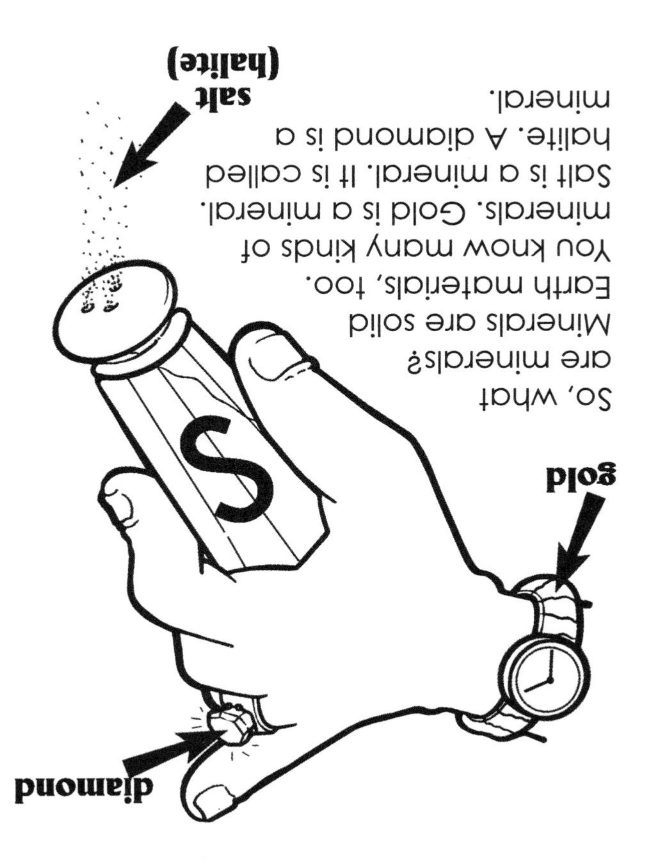


3

A mineral has the same material all the way through. Rocks can be made of many minerals. The tiny specks of color in rocks are the different minerals.

4

**Topic**
Minerals

**Key Question**
How can we identify common minerals using some of their observable attributes?

**Learning Goals**
Students will:
- observe minerals,
- explore some properties used for mineral identification, and
- identify minerals using an identification key.

**Guiding Documents**
*Project 2061 Benchmark*
- *Rock is composed of different combinations of minerals. Smaller rocks come from the breakage and weathering of bedrock and larger rocks. Soil is made partly from weathered rock, partly from plant remains—and also contains many living organisms.*

*NRC Standard*
- *Simple instruments, such as magnifiers, thermometers, and rulers, provide more information than scientists obtain using only their senses.*

**Science**
Earth science
 minerals

**Integrated Processes**
Observing
Predicting
Collecting and recording data
Comparing and contrasting
Generalizing

**Materials**
Sticky dots, quarter-inch
Mineral set for *Part Two* (see *Management 1)*
*Mineral Management* journal (see *Management 6)*

*For each station in Part One:*
**Quartz Station**
 quartz
 halite
 steel file
 10-penny nail
 hand lens

**Graphite Station**
 graphite
 hand lens

**Hematite Station**
 hematite
 unglazed white tile
 hand lens

**Mica Station**
 mica
 hand lens

**Pyrite Station**
 pyrite
 unglazed white tile
 hand lens

**Gypsum Station**
 gypsum
 hand lens

**Feldspar Station**
 feldspar
 hand lens

**Magnetite Station**
 magnetite
 halite
 unglazed white tile
 ring magnet
 hand lens

**Calcite Station**
 calcite
 halite
 quartz
 vinegar (see *Management 4)*
 hand lens
 eyedropper

 

**Hornblende Station**
hornblende
unglazed white tile
hand lens

**Galena Station**
galena
halite
unglazed white tile
hand lens

*For Mineral Identification Stations in Part Two:*
**Mineral Identification Station One**
Mystery Minerals 1, 2, 3 (see *Management 5)*
hand lens
vinegar
eyedropper
10-penny nail
steel file

**Mineral Identification Station Two**
Mystery Minerals 4, 5, 6 (see *Management 5)*
hand lens

**Mineral Identification Station Three**
Mystery Minerals 7, 8, 9 (see *Management 5)*
hand lens
unglazed white tile

**Mineral Identification Station Four**
Mystery Minerals 10, 11, 12 (see *Management 5)*
hand lens
unglazed white tile
ring magnet

**Background Information**

A mineral is a naturally occurring inorganic element or compound with a specific chemical composition, a unique crystal structure, and special chemical properties. Most minerals have key characteristics that help in their identification. Some of the most common characteristics include color, shape of the crystal, hardness, luster, streak, cleavage/fracture, and specific gravity.

Color is simply the basic color of the mineral, but in many cases it is the least useful attribute for identification.

The physical hardness of a mineral is usually measured according to Mohs' Scale of Mineral Hardness. Each mineral has a hardness number going from very soft minerals like talc to very hard minerals such as diamonds. There is a simplified field scale that uses everyday objects.

Luster describes the way a mineral's surface interacts with light and can range from dull to glassy.

Streak refers to the color of the powder a mineral leaves after rubbing it on an unglazed porcelain streak plate.

Cleavage and fracture are often examined together. Cleavage describes the way a mineral may come apart or cleave in different ways. Fracture describes how a mineral breaks when broken contrary to its natural cleavage planes.

Specific gravity is the ratio of the mineral weight to the weight of an equal volume of water. This is often described as the density of the mineral. Other properties used for mineral identification include magnetism and reactivity to dilute acids.

**Management**

1. Mineral sets (item number 4108M), hand lenses (item number 1977), ring magnets (item number 1971), and eyedroppers (item number 1946) are available from AIMS.
2. For *Part One,* place a copy of the *Mineral Management* cards, along with the materials for each station, at different locations around the room. Determine a pattern of rotation for the students to follow so they can complete each station in an orderly fashion.
3. For *Part Two,* you will need to set up the four mystery stations (see *Materials).* Students will use the information recorded in their journals to help them identify the mystery minerals. Plan a rotation pattern for these stations as well. If desired, you can label these stations 1, 2, 3, 4 using index cards.
4. You will need to boil 8 ounces of vinegar for about 10 minutes to create a stronger acid solution to test the calcite for reaction to an acid. Calcite will bubble when acid is dropped on its surface. Students will be able to better see the reaction if they use the hand lens. The reaction will not produce a frothy bubbling. Instead, students should see bubbles on the surface of the rock. The tiny bubbles will look much like the bubbles seen on the side of a glass containing clear soda pop. If the acid is not strong enough to create bubbles, explain to the students that a stronger acid is needed but it would be dangerous for them to use.
5. For *Part Two,* you will need to cover the identification numbers of the minerals. Quarter-inch sticky dots can be used for this purpose. Before putting the sticky dots on the minerals, assign them the following numbers: Halite–1, Calcite–2, Quartz–3, Mica–4, Gypsum–5, Feldspar–6, Galena–7, Pyrite–8, Hornblende–9, Magnetite–10, Hematite–11, Graphite–12.
6. Each student needs a *Mineral Management* journal to record observations. Pages can be copied front to back, cut apart, and stapled along the left side to make the journal.

 

## Procedure

*Part One*

1. Ask the *Key Question* and state the *Learning Goals.*
2. Distribute and read the *Mineral Management* rubber band book. Point out that most minerals need multiple tests for identification.
3. Tell the students that most minerals have key characteristics that help in their identification. Halite is a good example. Some key characteristics of halite are its cubed-shaped crystals, its lack of color, and how easily it can be scratched. Knowing these characteristics helps identify halite.
4. Point out the location of the mineral observation cards. Tell the students that they will be making some observations about other minerals. They will need to keep a written record of what they find out at each station. Distribute the journals. Direct students to answer the questions that are on the pages. Remind them to clean the area before moving to the next station.
5. Allow time for the students to complete each station.

*Part Two*

1. Ask the *Key Question* and state the *Learning Goals.*
2. Tell the students that they will be using their written records to identify some minerals based on attributes. Point out the location of the four mystery stations and explain to the students that they can use any materials at the individual stations to help identify the mystery minerals.
3. Allow time for each group of students to rotate through the four stations.
4. Discuss the results.

## Connecting Learning

1. What are some of the attributes you observed about the minerals?
2. Why do you think color may not be the best way to identify minerals?
3. What mineral identification station was the most difficult? Why?
4. What are some unique attributes of minerals?
5. How did the hand lens help with observing attributes of the minerals?
6. What are you wondering now?

## Key Question

How can we identify common minerals using some of their observable attributes?

## Learning Goals

**Students will:**

- observe minerals,
- explore some properties used for mineral identification, and
- identify minerals using an identification key.

7

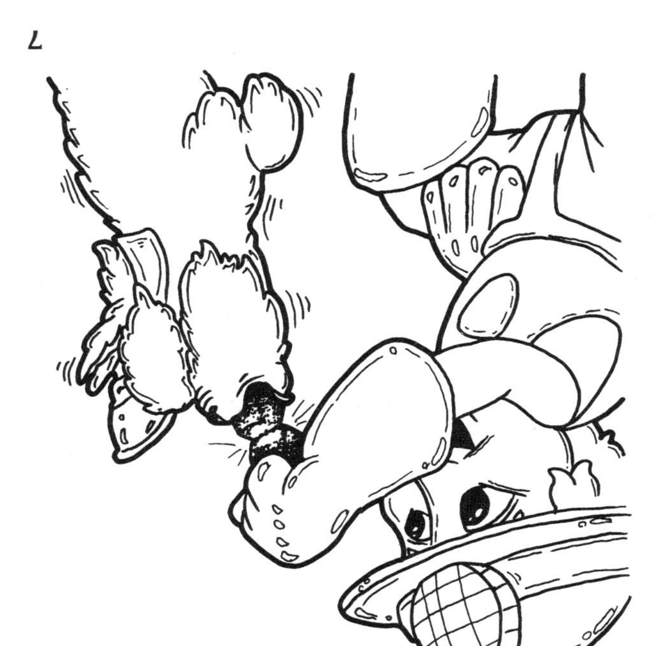

Some minerals have special properties. Some are magnetic. Others bubble when an acid is dropped on them.

2

Rocks are made of minerals. They occur naturally on the Earth. There are tests you can use to identify minerals.

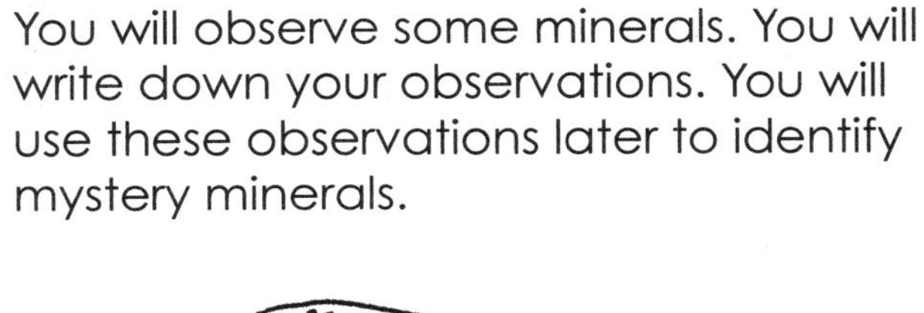

You will observe some minerals. You will write down your observations. You will use these observations later to identify mystery minerals.

8

1

 

Color is one property of minerals. Some mineral colors are brown, black, and yellow. Some look almost clear.

3

4

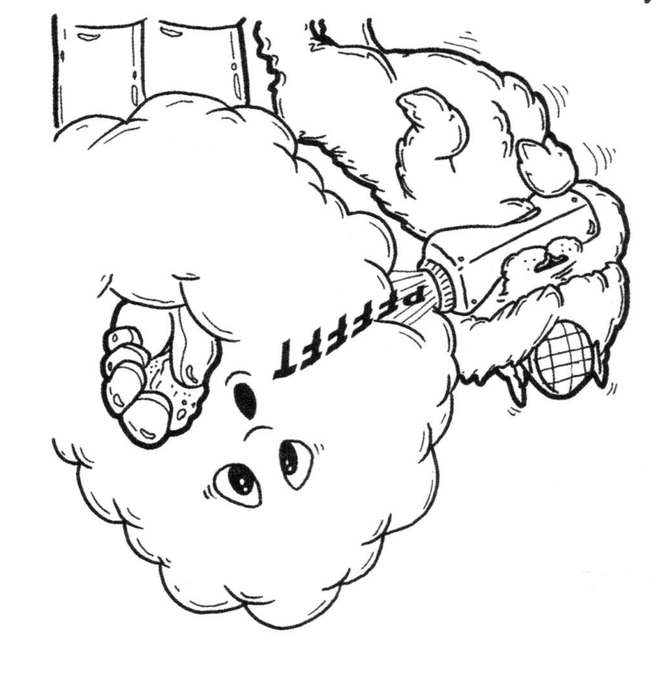

Minerals can be soft. Talc is very soft. It is used to make baby powder. Other minerals, like diamonds, are hard.

5

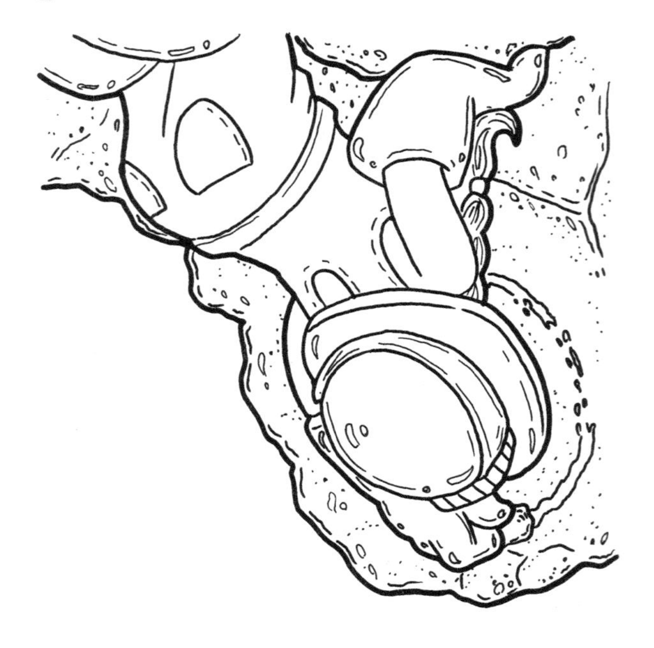

Some minerals can be used to draw a colored line on a tile. Others will not leave a color at all.

The surfaces of some minerals are very shiny. Some are very dull. Some have flat surfaces. Some have very rough surfaces. Some have both types. Some are very heavy. Some are very light.

6

MINERAL MANAGEMENT
QUARTZ STATION
1. List attributes of quartz.
2. What type of crystal shape can you observe?
3. Try scratching quartz and halite with a file and a nail. What happened?
4. What two attributes would help you identify quartz?

 

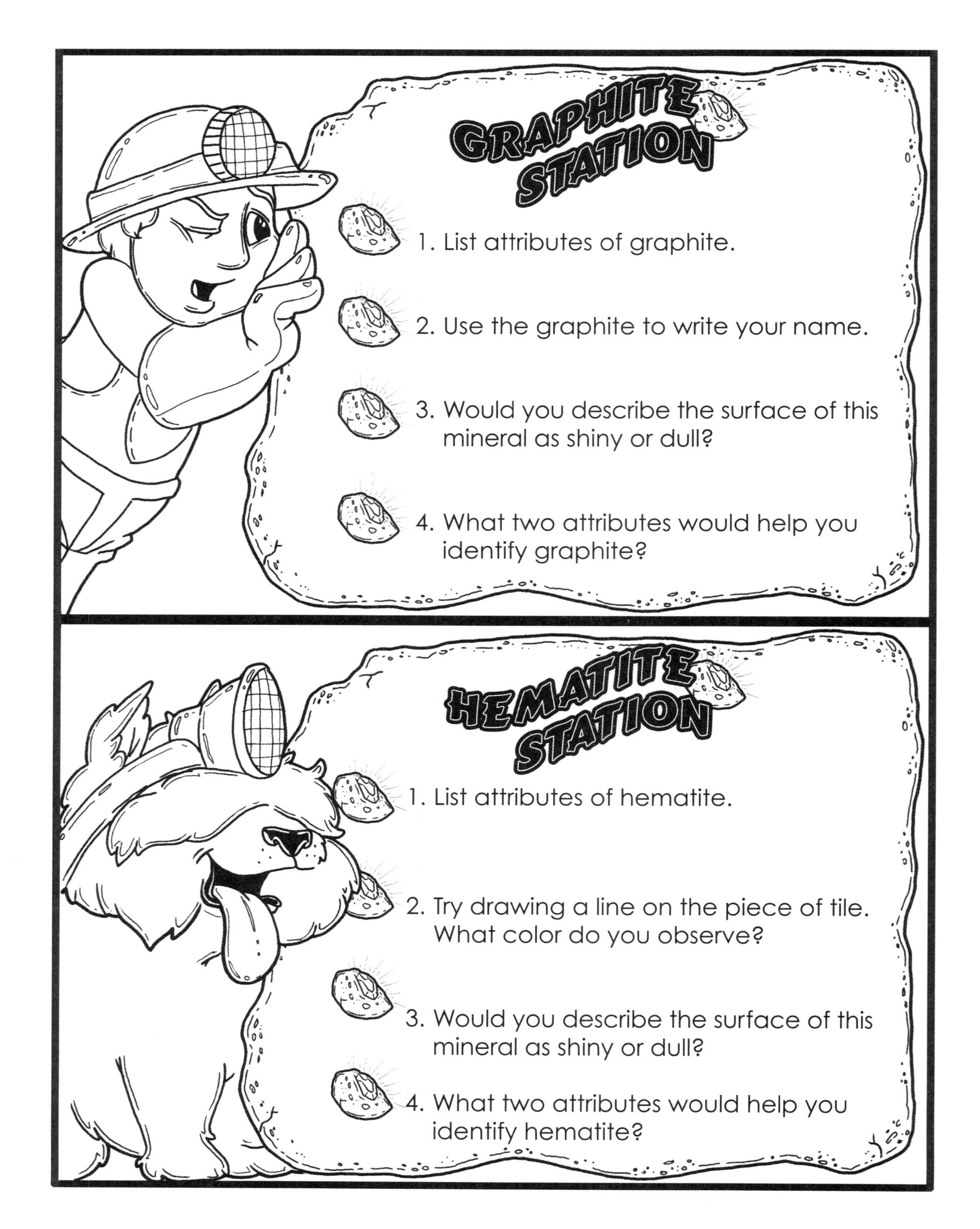
GRAPHITE STATION
1. List attributes of graphite.
2. Use the graphite to write your name.
3. Would you describe the surface of this mineral as shiny or dull?
4. What two attributes would help you identify graphite?
HEMATITE STATION
1. List attributes of hematite.
2. Try drawing a line on the piece of tile. What color do you observe?
3. Would you describe the surface of this mineral as shiny or dull?
4. What two attributes would help you identify hematite?

MICA STATION
1. List attributes of mica.
2. Would you describe the surface of this mineral as shiny or dull?
3. Notice that the mineral looks like it is in layers or sheets.
4. What two attributes would help you identify mica?
PYRITE STATION
1. List attributes of pyrite.
2. Draw a line on the piece of tile. What color do you observe?
3. Why do you think this is sometimes called fool's gold?
4. What two attributes would help you identify pyrite?

GYPSUM STATION
1. List attributes of gypsum.
2. Notice the columns that seem to run along in one direction.
3. Use your fingernail to try to scratch the surface of the gypsum. Would you describe this mineral as hard or soft?
4. What two attributes would help you identify gypsum?
FELDSPAR STATION
1. List attributes of feldspar.
2. Notice that some areas of the feldspar are flat and some are jagged.
3. Would you describe the surface of this mineral as dull, shiny, or both?
4. What two attributes would help you identify feldspar?

 

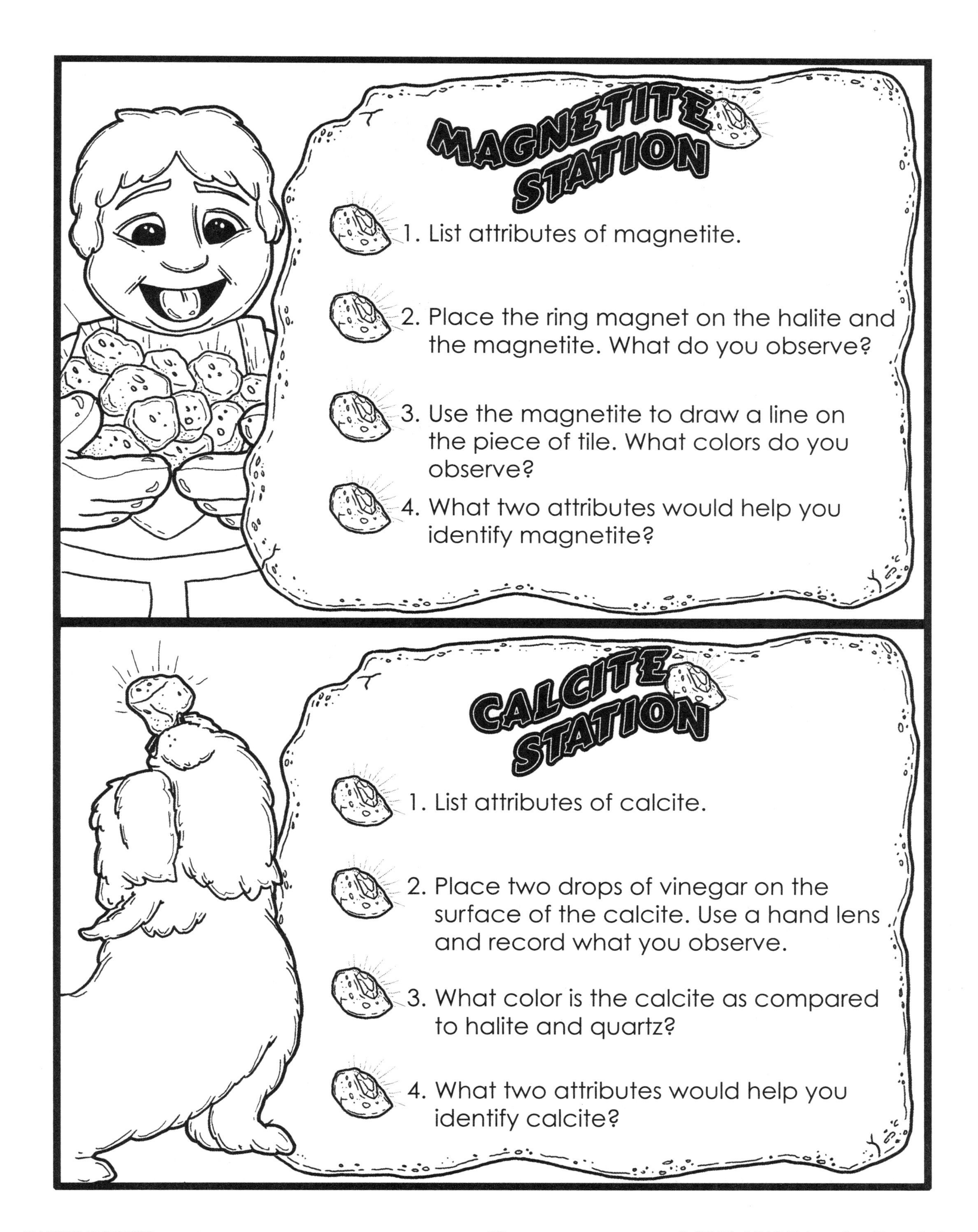
MAGNETITE STATION
1. List attributes of magnetite.
2. Place the ring magnet on the halite and the magnetite. What do you observe?
3. Use the magnetite to draw a line on the piece of tile. What colors do you observe?
4. What two attributes would help you identify magnetite?
CALCITE STATION
1. List attributes of calcite.
2. Place two drops of vinegar on the surface of the calcite. Use a hand lens and record what you observe.
3. What color is the calcite as compared to halite and quartz?
4. What two attributes would help you identify calcite?

HORNBLENDE STATION
1. List attributes of hornblende.
2. Draw a line on the piece of tile. What color do you observe?
3. How would you describe the surface of this mineral?
4. What two attributes would help you identify hornblende?
GALENA STATION
1. List attributes of galena.
2. Draw a line on the piece of tile. What color do you see?
3. Pick up a piece of halite and piece of galena. What do you observe?
4. What two attributes would help you identify galena?

MINERAL MANAGEMENT
MYSTERY MINERALS
STATION 1
NAME THE MINERALS
1.
2.
3.
STATION 2
NAME THE MINERALS
1.
2.
3.
STATION 3
NAME THE MINERALS
1.
2.
3.
STATION 4
NAME THE MINERALS
1.
2.
3.

CONNECTING LEARNING

## Connecting Learning

1. What are some of the attributes you observed about the minerals?
2. Why do you think color may not be the best way to identify minerals?
3. What mineral identification station was the most difficult? Why?
4. What are some unique attributes of minerals?
5. How did the hand lens help with observing attributes of the minerals?
6. What are you wondering now?

**Topic**
Rocks and minerals

**Key Question**
What can you learn about these two samples by making careful observations?

**Learning Goals**
Students will:
- make careful observations about two samples of Earth's materials, and
- compare and contrast rocks and minerals.

**Guiding Documents**
*Project 2061 Benchmark*
- *Rock is composed of different combinations of minerals. Smaller rocks come from the breakage and weathering of bedrock and larger rocks. Soil is made partly from weathered rock, partly from plant remains—and also contains many living organisms.*

*NRC Standard*
- *Some changes in the solid earth can be described as the "rock cycle." Old rocks at the earth's surface weather, forming sediments that are buried, then compacted, heated, and often recrystallized into new rock. Eventually, those new rocks may be brought to the surface by the forces that drive plate motions, and the rock cycle continues.*

**Science**
Earth science
rocks and minerals

**Integrated Processes**
Observing
Comparing and contrasting
Relating
Communicating
Collecting and recording data
Interpreting data
Inferring

**Materials**
*For each student group:*
granite sample
quartz sample
hand lens

**Background Information**
Minerals are naturally-occurring inorganic substances. Inorganic means they are made up of things that are not alive. Minerals have very specific chemical and physical properties. Some minerals consist of only one element, but most are compounds. Minerals are usually identified by their physical properties. Some of these properties are hardness and specific gravity. Quartz, feldspar, and mica are examples of minerals.

Rocks are natural combinations of one or more minerals. Rocks are classified based on the conditions under which they formed.

*Sedimentary Rocks*
Sedimentary rocks are formed by the action of wind, water, snow, or organisms. They cover about three-fourths of the Earth's surface. Most are deposited as sediments on the bottom of rivers, lakes, and seas. Many sedimentary rocks have been moved by water, wind, waves, currents, ice, or gravity. The most common sedimentary rocks are sandstone, limestone, conglomerates, and shale.

*Igneous Rocks*
Igneous rocks are formed at very high temperatures. They come from magmas, which are molten mixtures of minerals. These magmas come from deep within the Earth. If they cool off below the surface, they form intrusive rocks, which may later be uncovered through the process of erosion. When magmas reach the surface and cool, they form extrusive rocks. Granite and basalt are examples of igneous rocks.

*Metamorphic Rocks*
Metamorphic rocks are those that have been changed from some other type of rock. They change as a result of heat, pressure, or chemical action. All kinds of rocks can be changed. The result is a new crystal structure, the formation of new minerals, or a change in the rock's texture. Slate was once shale. Marble comes from limestone. Granite can become gneiss.

 

**Management**

1. Each student group needs a piece of granite and a piece of quartz.
2. Rock and mineral sets (item number 4108) and hand lenses (item number 1977) are available from AIMS.

**Procedure**

1. Ask the *Key Question* and state the *Learning Goals.*
2. Distribute the granite and quartz. Encourage the students to make as many observations as they can about the two.
3. Direct a discussion centered on the students' observations. Point out to them that one of the many observations they could have made is that one sample appears to be composed of one substance and one is composed of more than one substance.
4. Distribute the information on rocks and minerals. Help the students read the information and respond to the questions presented. Do this as a class because of the difficulty of the reading passage.

**Connecting Learning**

1. How are rocks and minerals alike and different?
2. What are the three types of rocks?
3. Which type of observation—sketch or written—do you think gave the best observations about the two rocks? Why?
4. How did the reading passage help you learn more about rocks and minerals?
5. What are you wondering now?

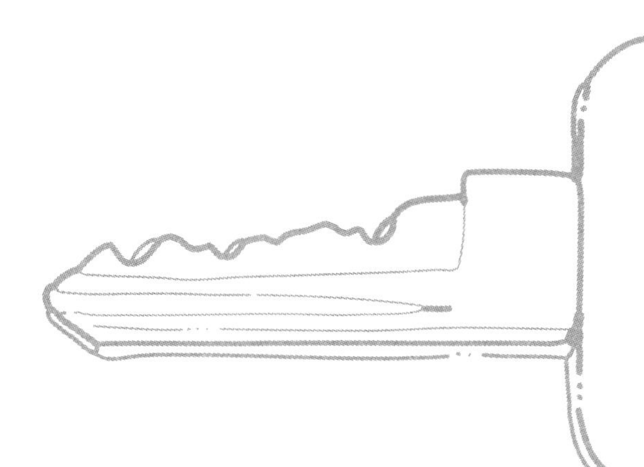

## Key Question

What can you learn about these two samples by making careful observations?

## Learning Goals

**Students will:**

- make careful observations about two samples of Earth's materials, and
- compare and contrast rocks and minerals.

Make a record of observations. Include a sketch as well as written descriptions as to how the samples are alike and different.

## Observations

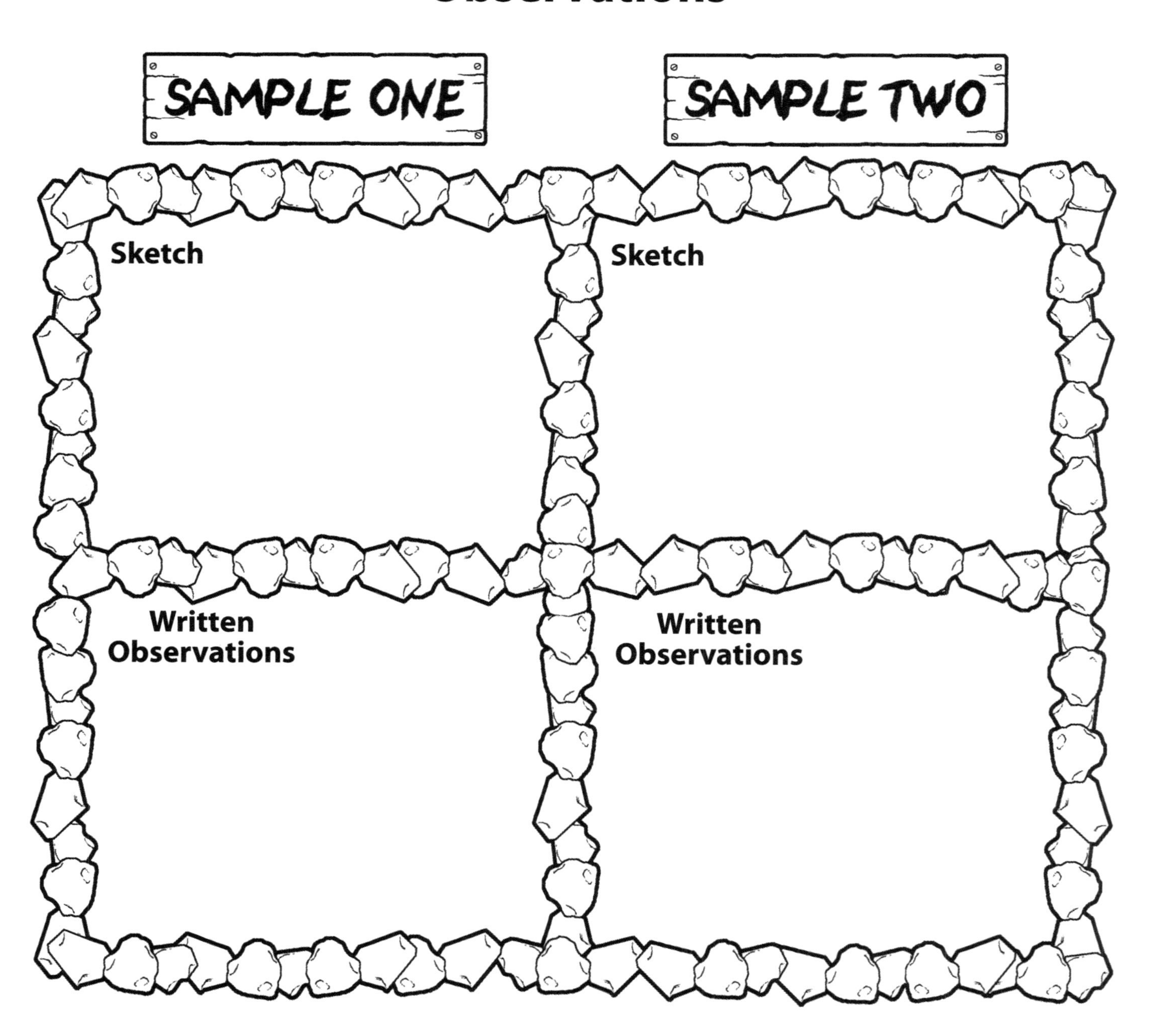


 

# Just Between the Two of Them

## Rocks and Minerals

Geologists define minerals as naturally-occurring inorganic substances. Inorganic means minerals are made up of things that are not alive. Some minerals can be made up of only one material, but most are made up of more. Minerals are usually identified based on their physical properties. Some of these properties are hardness and the shape of the crystals found in the mineral. Quartz, feldspar, and mica are examples of minerals.

Rocks are natural combinations of one or more minerals. Rocks are classified based on the conditions under which they formed. The three different types of rocks are sedimentary, igneous, and metamorphic. Sedimentary rocks are formed from other rocks that have been broken down. Igneous rocks are formed from melted rocks. Metamorphic rocks are rocks that form when other rocks undergo changes through heat and pressure. Granite and limestone are examples of rocks.

Answer these questions in your science journal.

1. What is the difference between rocks and minerals?

2. What would an organic substance be?

3. What are the three types of rocks?

4. What do you think the word *metamorphic* means?

5. What do you think was the author's purpose in writing this passage?

CONNECTING LEARNING

## Connecting Learning

1. How are rocks and minerals alike and different?
2. What are the three types of rocks?
3. Which type of observation—sketch or written—do you think gave the best observations about the two rocks? Why?
4. How did the reading passage help you learn more about rocks and minerals?
5. What are you wondering now?

 

Living materials in the soil might be plant roots or tiny animals.

Once-living things are dead plants and animals. These decay and add nutrients to the soil. The decaying matter is called humus.



Much of the land surface of the Earth is covered with soil. The soil can be a few inches deep or it can go down several feet.

Soil is important to life on Earth. It supports the plants and animals on which humans and other living things depend.

Soil is a mixture. It is made up of living, once-living, and nonliving materials.

Nonliving materials in soil are sand, silt, and clay.

3

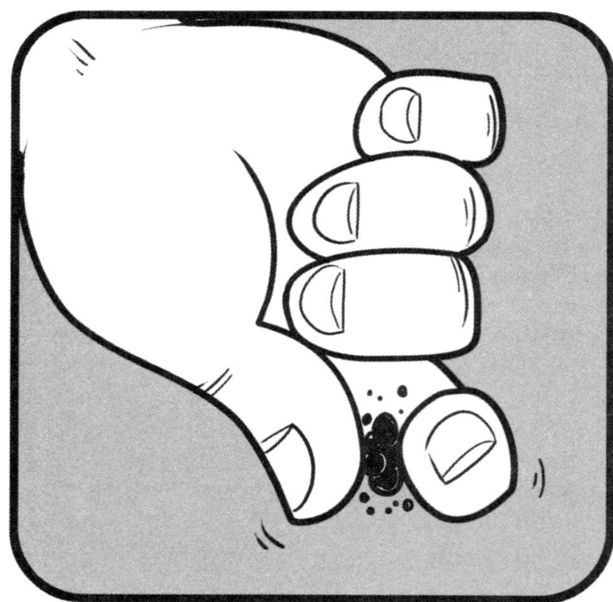

The texture of soil depends on the sizes of these nonliving materials. Rub a moist piece of soil between your fingers. If it feels slick, it contains a lot of clay. Clay is made up of very small particles.

4

If it feels gritty, it contains a lot of sand. Particles of sand are the biggest of the three.

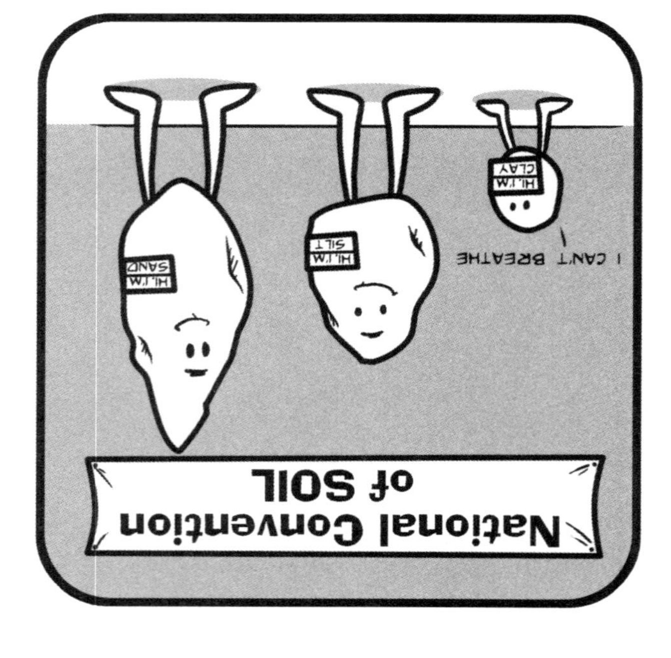


5

Silt feels somewhere between the two. Silt particles are bigger than clay but smaller than sand.

6

**Topic**
Soil composition

**Key Question**
How do soils from different sources compare?

**Learning Goal**
Students will explore the components of different soil samples.

**Guiding Documents**
*Project 2061 Benchmark*
- *Rock is composed of different combinations of minerals. Smaller rocks come from the breakage and weathering of bedrock and larger rocks. Soil is made partly from weathered rock, partly from plant remains—and also contains many living organisms.*

*NRC Standard*
- *Soils have properties of color and texture, capacity to retain water, and ability to support the growth of many kinds of plants, including those in our food supply.*

**Math**
Measurement
mass

**Science**
Earth science
soil

**Integrated Processes**
Observing
Comparing and contrasting
Collecting and recording data

**Materials**
*For each group:*
hand lens
balance
metric masses
3 soil samples (see *Management 2)*
tablespoon measure
toothpicks
cup of water
wax paper (see *Management 4)*
soil samples pages

*For each student:*
observation journal

**Background Information**
Soil contains a variety of materials including gravel, small rock particles, sand, decaying organic matter, clay, and so on. The kind and quantity of materials determine the type and texture of the soil. They also determine its ability to retain water, support plant life, etc. Soils from locations only feet apart can have vastly different compositions. This activity provides students the opportunity to explore soils from different locations and compare them in a variety of ways—texture, color, ability to retain water, and so on.

**Management**
1. Students should work in groups of three to five.
2. Several days prior to doing this activity, have each student bring in a small (½ cup or less) sample of soil in a resealable plastic bag. Have them identify the soil by writing the location from which it was taken on the plastic bag using a permanent marker (i.e., playground, garden, beach, riverbank, park, etc.).
3. You may want to collect several soil samples yourself to ensure a wide variety in the kinds of soil to be studied.
4. Give each group a small cup of water and wax paper so that they can test what happens when they add water to each soil sample. The wax paper only needs to be large enough to hold a small sample of soil—one tablespoon or less.
5. To make the observation journals, copy the pages and cut them along the dashed lines. Arrange the pages in order and staple along the left edge. Each student needs his or her own journal.
6. Hand lenses (item number 1977), balances (item number 1917), and metric masses (item number 1923) are available from AIMS.

 

**Procedure**

1. Set out the collection of soil samples on a table so that all can be seen.
2. Divide students into their groups and have one representative from each group come to the table and select three soil samples. Encourage them to select samples that appear to be different from each other.
3. Distribute the materials to each group and the journals to each student. Read through the journals together so that students understand the things they will be doing to explore their soil samples.
4. Provide time for students to work through the different experiments and record their observations and discoveries.
5. Discuss what students discovered and how the soils compare.

**Connecting Learning**

1. What colors were your soil samples? Were they one color or many colors?
2. Describe the texture of your soil samples. How do they feel?
3. What happened to your soil samples when you sprinkled them with water? Did they hold their shape? Did they absorb the water?
4. Which of your soil samples was the lightest? What was its mass? Which was the heaviest? What was its mass?
5. What do you think makes some samples heavier than others?
6. How were the soils you sampled alike? How were they different?
7. Were there any samples from different locations that turned out to be very similar? What might be some reasons for this?
8. Were there any samples from the same location that turned out to be very different? What might be some reasons for this?
9. What are you wondering now?

**Extension**

Study different soil layers. Compare soil from the top of the ground with soil from one and two feet under the same spot.

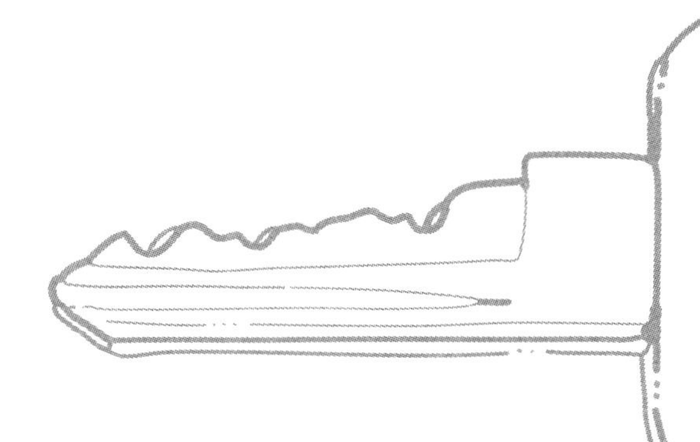

## Key Question

How do soils from different sources compare?

## Learning Goal

**Students will:**

explore the components of different soil samples.

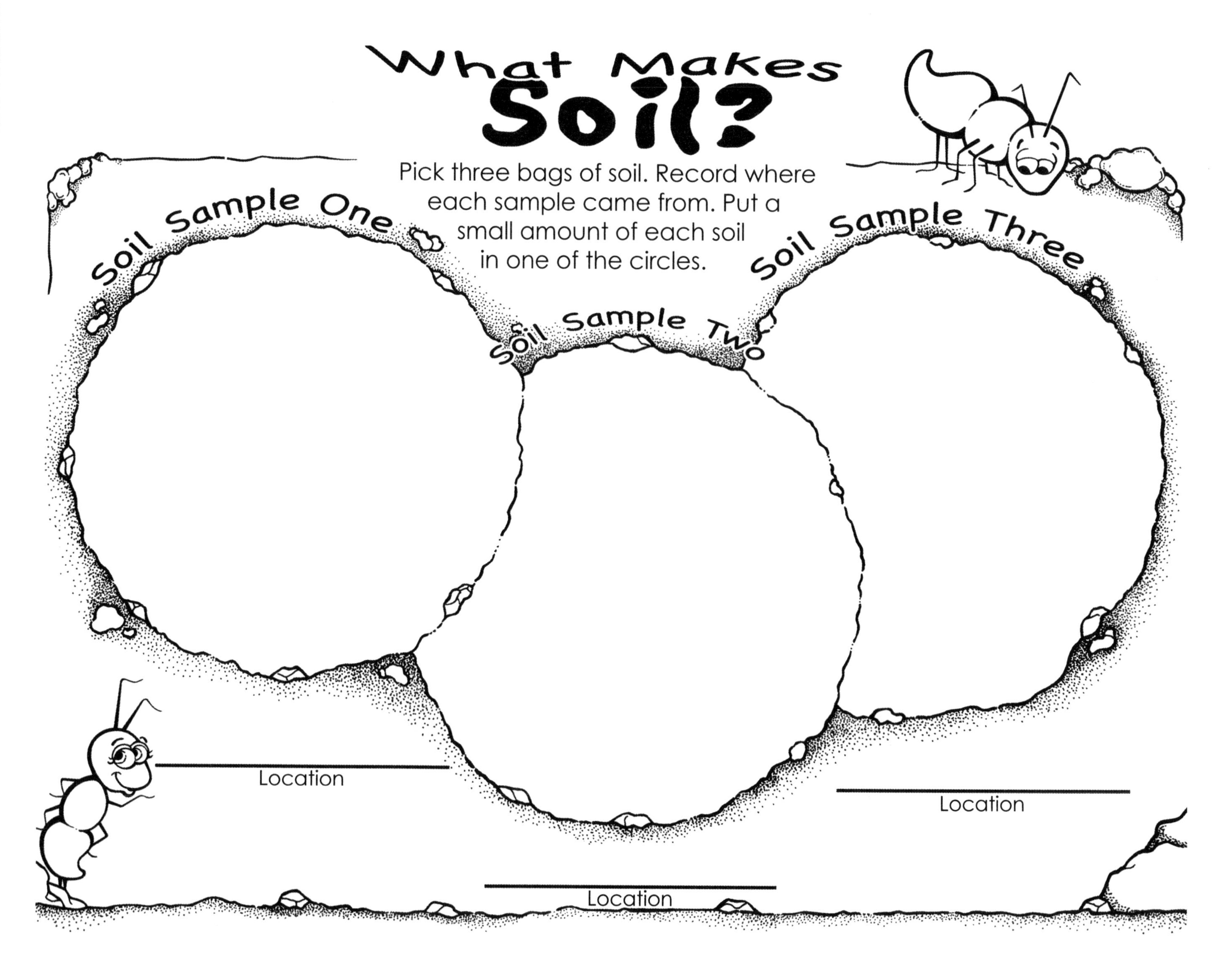

What Makes Soil?
Pick three bags of soil. Record where each sample came from. Put a small amount of each soil in one of the circles.
Soil Sample One
Soil Sample Two
Soil Sample Three
Location
Location
Location

## Observation Journal

1

Look carefully at your three soil samples. Use a hand lens. Separate the soil with a toothpick.

What color is the soil?

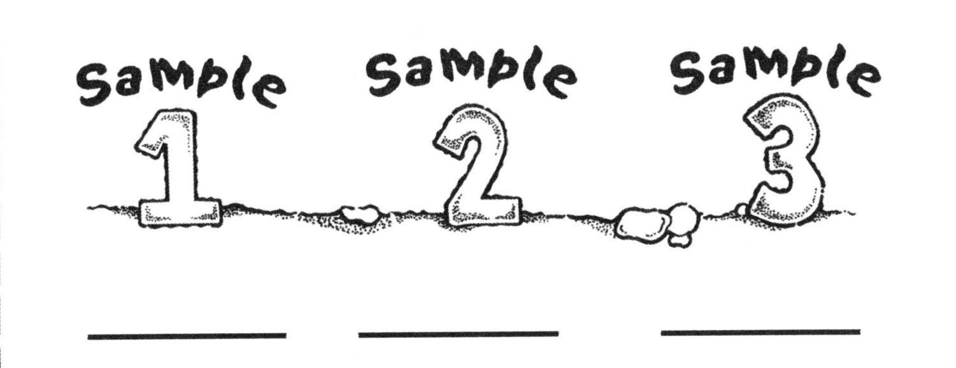


Are there any bits of rock?

Yes No Yes No Yes No

2

Rub a pinch of each soil between your thumb and forefinger.
Describe how it feels.
Sample 1
Sample 2
Sample 3
3
Hold each sample close to your ear and rub.
What do you hear?
If you hear a scratchy sound, it probably means that there is sand in your sample.
Sample 1
Sample 2
Sample 3
4

 

Put the soil on wax paper. Sprinkle a small amount of water on each sample.

Does the soil hold its shape?

Yes No     Yes No     Yes No

Can you form it into an object?

Yes No     Yes No     Yes No

5

Measure three tablespoons of soil into the balance. Guess what its mass is. Use the balance to find the actual mass.

**Sample 1**

________ g     ________ g
My guess     Actual mass

**Sample 2**

________ g     ________ g
My guess     Actual mass

**Sample 3**

________ g     ________ g
My guess     Actual mass

Order the soil samples from lightest to heaviest.

________ ________ ________
Light ⟶ Heavy

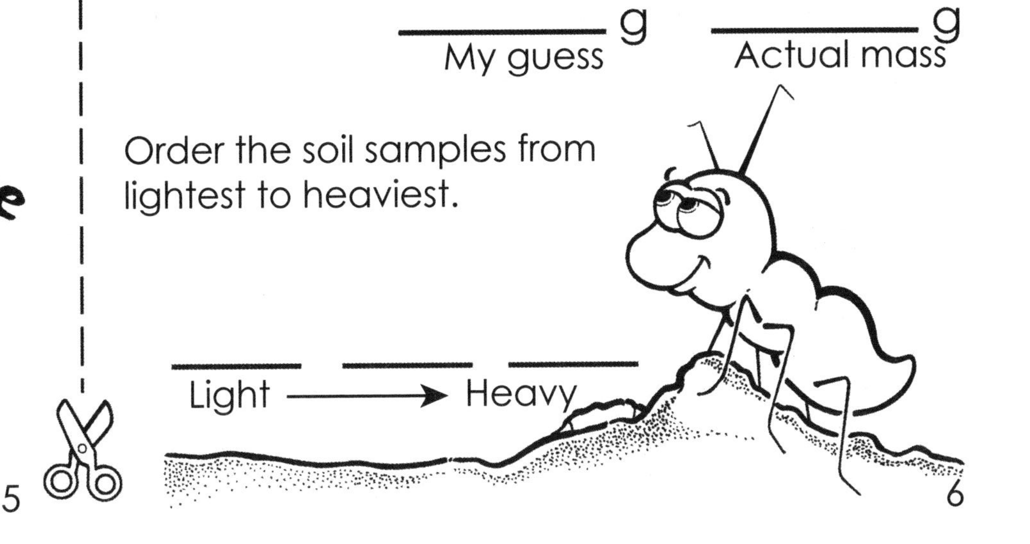

6

## Connecting Learning

1. What colors were your soil samples? Were they one color or many colors?

2. Describe the texture of your soil samples. How do they feel?

3. What happened to your soil samples when you sprinkled them with water? Did they hold their shape? Did they absorb the water?

4. Which of your soil samples was the lightest? What was its mass? Which was the heaviest? What was its mass?

5. What do you think makes some samples heavier than others?

## Connecting Learning

6. How were the soils you sampled alike? How were they different?

7. Were there any samples from different locations that turned out to be very similar? What might be some reasons for this?

8. Were there any samples from the same location that turned out to be very different? What might be some reasons for this?

9. What are you wondering now?

1

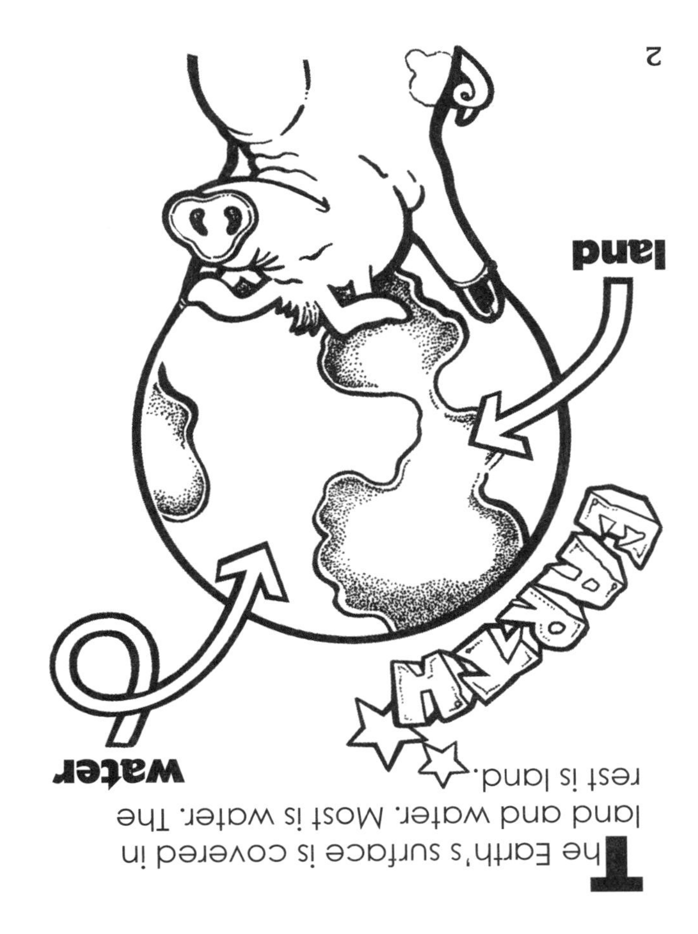


The Earth's surface is covered in land and water. Most is water. The rest is land.

2

Plants and animals also break down solid rock. Plant roots break rocks. Animals dig holes. People make paths.

7

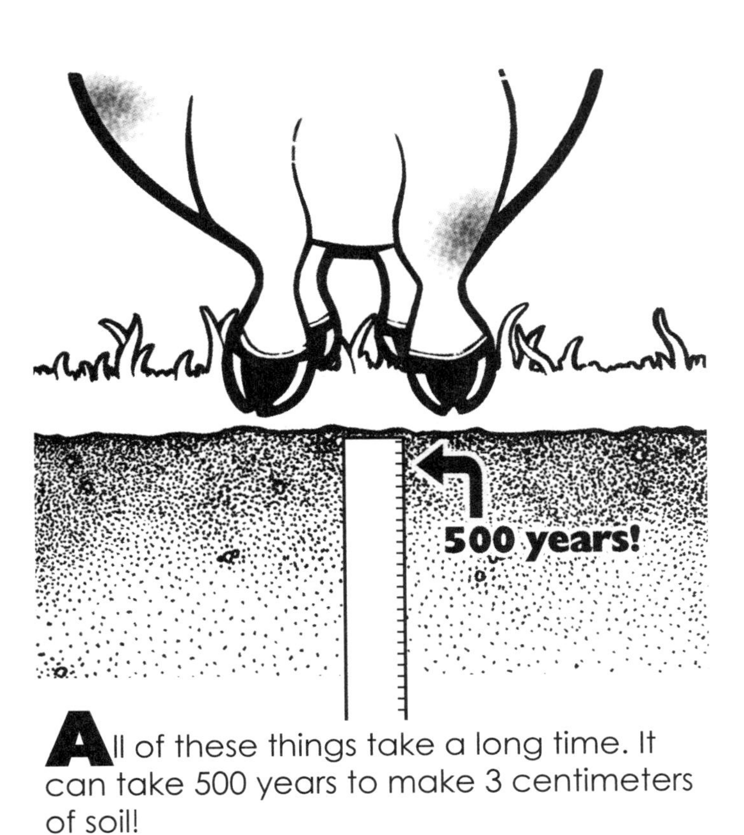


All of these things take a long time. It can take 500 years to make 3 centimeters of soil!

8

**M**ost land is covered with a thin layer of soil. Soil is usually the top 15-30 centimeters (6-12 inches).

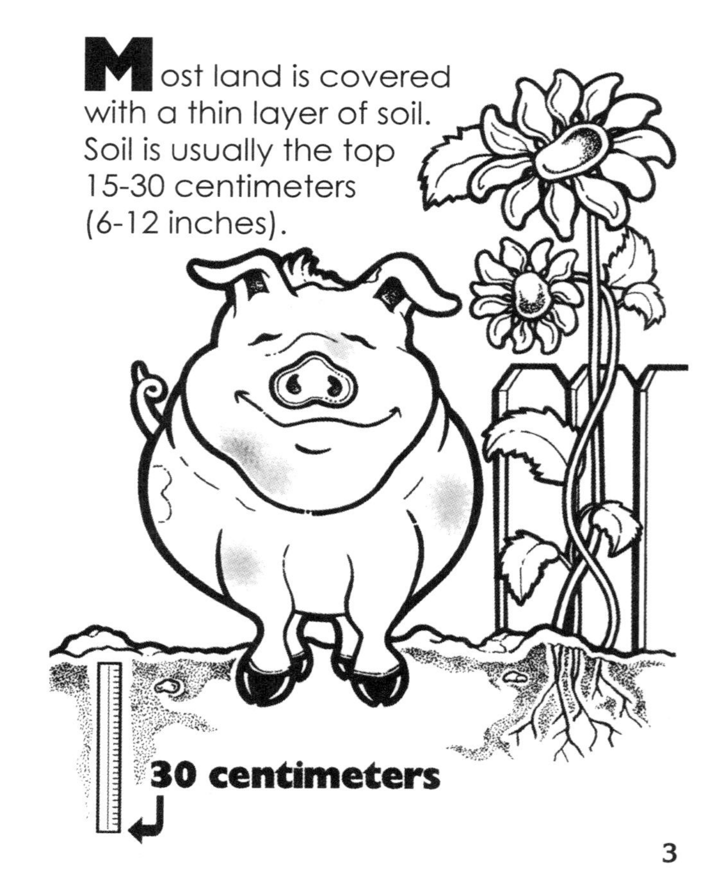


3

**S**oil can be made from many things. It often has small pieces of rock and bits of dead plants and animals. Sometimes it is very sandy. Other times, it has lots of clay. Soil from different places may be very different.

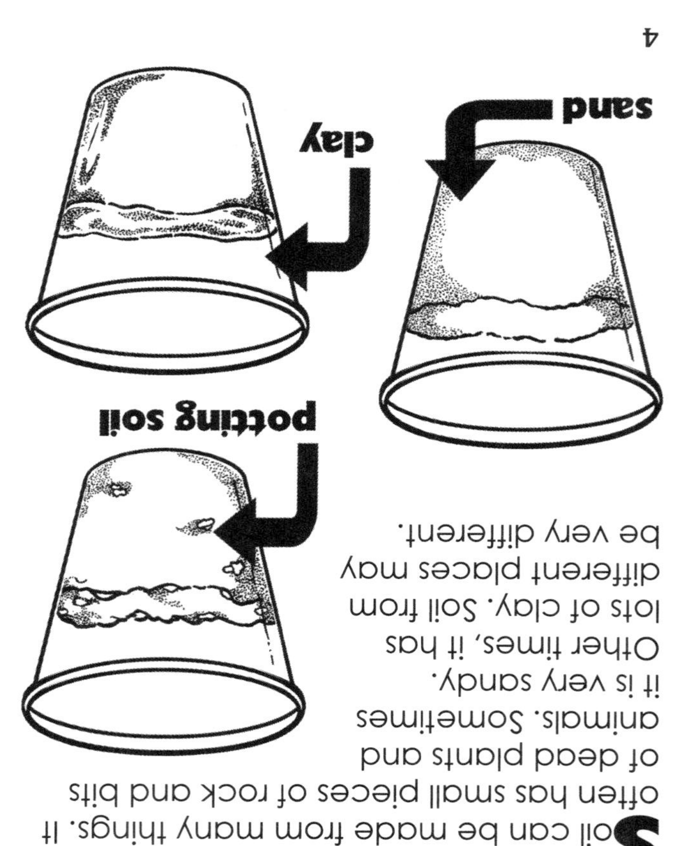


4

**S**oil forms in different ways. How does solid rock become soil?

5

**N**atural processes break down the rock. It can be broken down by:

- **wind**
- **rain**
- **heat/cold**
- **ice**
- **snow**

6

# Digging Into Soil Layers

**Topic**
Soil layers

**Key Question**
What are the layers in soil?

**Learning Goals**
Students will:
- learn about bedrock, subsoil, and topsoil; and
- make an edible model showing the components of these layers.

**Guiding Documents**
*Project 2061 Benchmarks*
- *Rock is composed of different combinations of minerals. Smaller rocks come from the breakage and weathering of bedrock and larger rocks. Soil is made partly from weathered rock, partly from plant remains—and also contains many living organisms.*
- *A model of something is different from the real thing but can be used to learn something about the real thing.*

*NRC Standard*
- *Soils have properties of color and texture, capacity to retain water, and ability to support the growth of many kinds of plants, including those in our food supply.*

**Science**
Earth science
  soil layers

**Integrated Processes**
Observing
Comparing and contrasting
Identifying

**Materials**
Nut brittle
Nuts (see *Management 2)*
Flaked coconut
Sunflower seeds
Brown sugar
Plastic cups (see *Management 1)*
Plastic spoons
Crayons or colored pencils
*Modeling Soil Layers* rubber band book
Rubber bands, #19
Student page

**Background Information**
Soil is critical to life on Earth. It covers much of the land surface and provides the nutrients and support necessary for plants to develop and grow. Soil is organized in layers, which scientists call horizons. Soil scientists have very specific and detailed ways in which they identify these horizons, and the designations can vary from country to country. For the purposes of this activity, the horizons (layers) of soil are simplified to bedrock, subsoil, and topsoil.

Topsoil is found on the top few inches to one foot of the soil. It is rich in humus and contains plant and animal life. Beneath topsoil is the subsoil. Here the minerals needed by plants can be found. This layer also contains a lot of clay. The bottom-most layer is called bedrock. This is the solid rock beneath all the soil. In places where hillsides have been cut away to make roads, it is often possible to see the soil layers.

**Management**
1. The cups used for the student models need to be transparent or translucent so that students can see the layers of their model soil. The 3 ½-oz size is recommended to cut down on the quantity of food materials needed. For the distribution of the food materials, use larger cups (9- or 10-oz) filled with enough materials for small groups of students.
2. Be aware of any nut allergies your students may have before doing this activity and make the necessary adjustments. Any kind of nuts and nut brittle can be used, but it should be consistent throughout each of the three layers. That is, if you use almond brittle, use almonds for the subsoil and chopped almonds for the topsoil. If nut brittles are not readily available, toffee candy bars can be used.
3. If students will be eating the models, be sure that they wash their hands thoroughly with soap and warm water before making the models. They should use spoons to put the food into their individual cups so that they are not touching anything that someone else will eat.

**Procedure**
1. Ask students what they know about soil and its layers. If necessary, ask them if they have seen the layers in road cuts. Record their thoughts on the board.
2. Explain that they will be studying three soil layers—bedrock, subsoil, and topsoil—and making a model of those layers.

 

3. Have students get into groups. Give each group cups filled with the food materials. Give each student a plastic spoon, a small plastic cup, and a copy of the rubber band book.
4. Instruct students to fold and assemble the rubber band book, then follow the directions on its pages to make their models. Be sure they use their spoons to scoop out food from the cups rather than their hands.
5. When all students have made their models, distribute the student page and crayons or colored pencils.
6. Have them draw a picture of their soil layers model in the cup on the page and identify each layer (bedrock, subsoil, topsoil).
7. Discuss how this model is like the real thing and how it is different.
8. Allow students to "dig in" and enjoy their edible models.

**Connecting Learning**

1. What are the layers of soil? [topsoil, subsoil, and bedrock]
2. What is the bedrock made of? [solid rock] How did we represent that in our model? [nut brittle]
3. What do you find in the subsoil? [clay, minerals] How did we represent that in our model? [nuts, with coconut for plant roots]
4. What things make up topsoil? [humus, clay, sand, silt, rock] How did we represent that in our model? [coconut, chopped nuts, sunflower seeds, and brown sugar]
5. Why do you think topsoil is the best for growing plants? [It has the organic material plants need and the animals that keep soil healthy.]
6. In what ways was our soil layer model like real soil? [It showed the layers and their relative order. It had different things for the different parts of soil.] In what ways was it different? [It's made of food. The relative depths of the layers are not shown. There are no living components. Not all the parts are represented. Etc.]
7. What are you wondering now?

**Extensions**

Have students brainstorm other food items to add to the soil to make the model more complete.

**Internet Connections**

*Discovery Education: The Dirt on Soil*
http://school.discoveryeducation.com/schooladventures/soil/index.html
Kid-friendly sections include information on the layers of soil, the organisms found in soil, and an interactive soil safari.

*Enchanted Learning: Soil Layers*
http://www.enchantedlearning.com/geology/soil/
Information on types of soil, soil formation, and soil horizons. Includes a diagram of soil layers.

 

# Digging Into Soil Layers

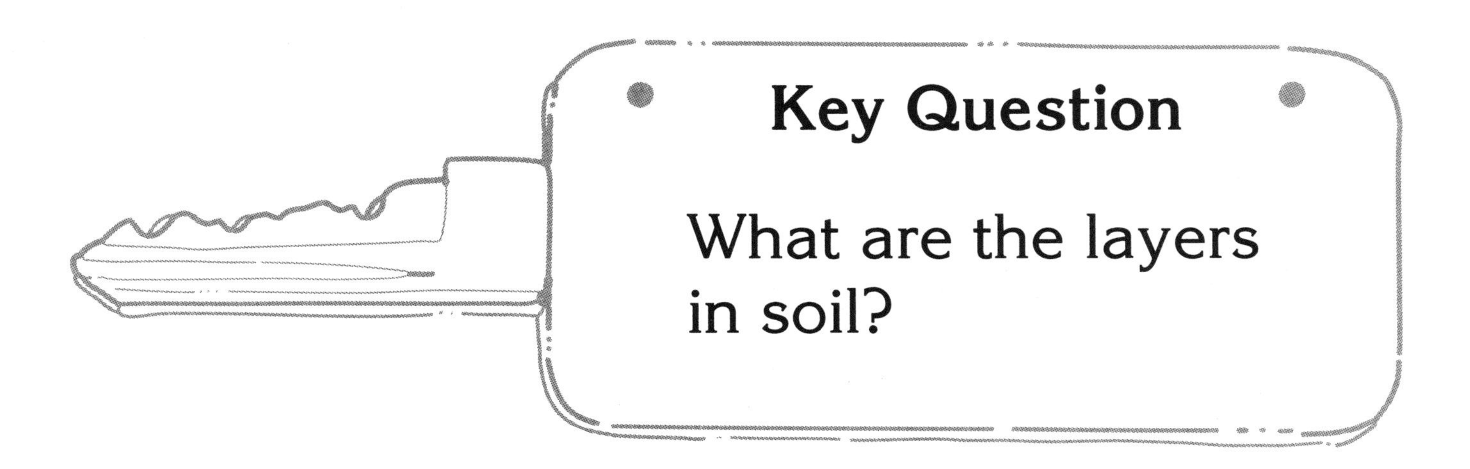


## Learning Goals

**Students will:**

- learn about bedrock, subsoil, and topsoil; and
- make an edible model showing the components of these layers.

# Digging Into Soil Layers

Soil has layers. You will be making a model of some of these layers. Read about the soil layers. As you read, make your model.

1

The deepest soil layer is bedrock. It is solid rock. There are no plant roots or animals here.

2

In our model, topsoil has four ingredients. Chopped nuts represent the clay and rocks. Coconut represents plant roots and animals. Sunflower seeds represent humus. Brown sugar represents silt and sand. Mix these four things. Put them on top of the nuts.

7

Compare your model to real soil. How are the layers alike? How are they different? Use the page to record your answers.

8

In our model, bedrock is nut brittle. Put a piece of nut brittle in the bottom of your cup.

3

Above bedrock is subsoil. It is a mixture of clay and minerals. Deep plant roots are found here.

4

In our model, subsoil is nuts. Put a layer of nuts on top of the nut brittle. Add a little coconut to represent plant roots.

5

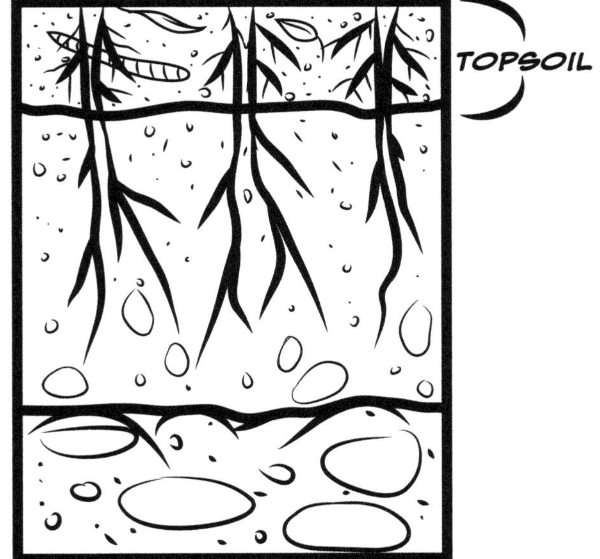


The top layer is called topsoil. It has rocks, clay, silt, and sand. It also contains humus. Humus is decaying plants and animals. Plant roots are found in topsoil. So are many animals.

6

 

# Digging Into Soil Layers

Draw a picture of your soil model. Label each layer.

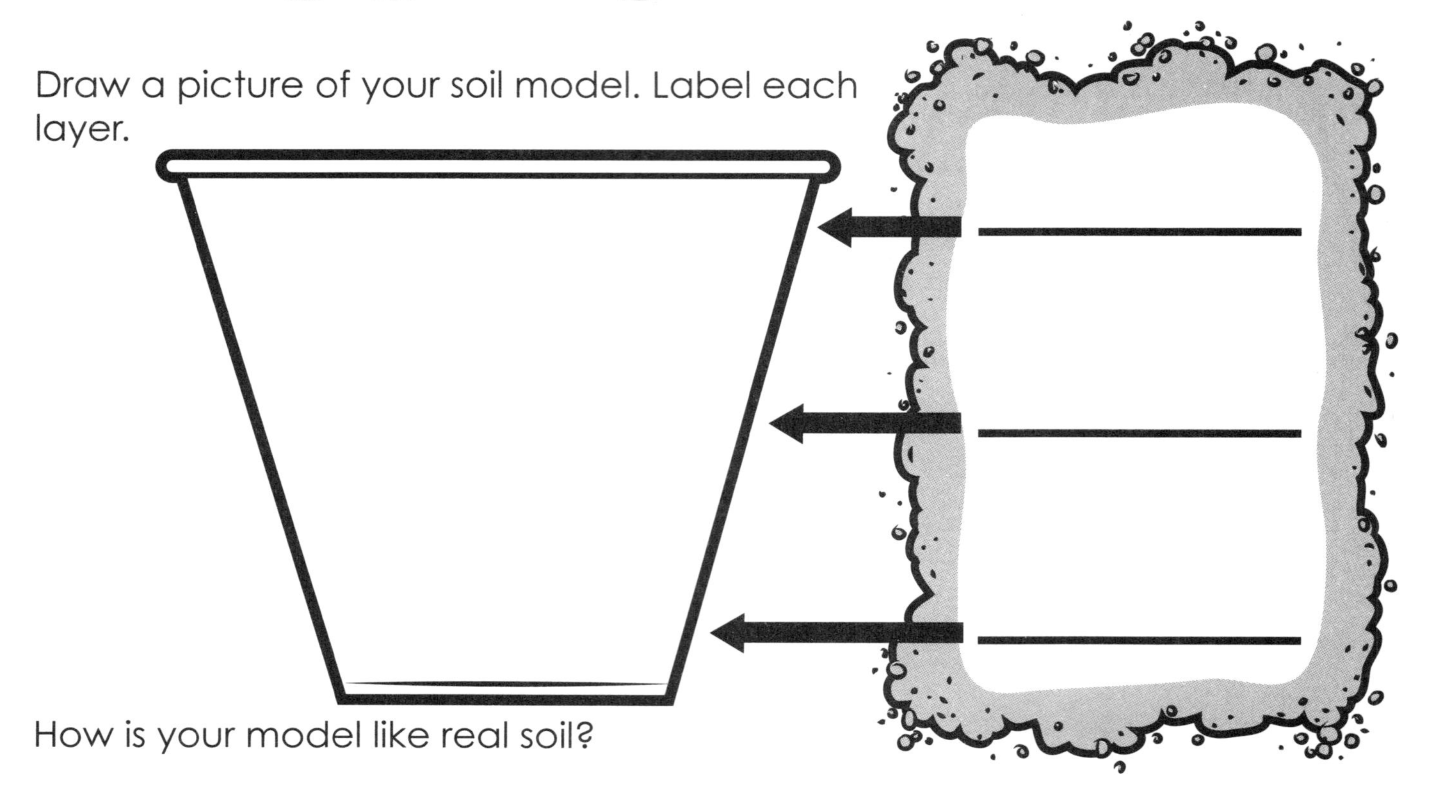

How is your model like real soil?

How is it different?

# Digging Into Soil Layers

CONNECTING LEARNING

## Connecting Learning

1. What are the layers of soil?
2. What is the bedrock made of? How did we represent that in our model?
3. What do you find in the subsoil? How did we represent that in our model?
4. What things make up topsoil? How did we represent that in our model?
5. Why do you think topsoil is the best for growing plants?
6. In what ways was our soil layer model like real soil? In what ways was it different?
7. What are you wondering now?

# A Record of Layers

**Topic**
Soil layers

**Key Question**
In what order are the layers of soil?

**Learning Goals**
Students will:
- read descriptions of the three layers of soil, and
- make a soil model by cutting out and ordering the layers.

**Guiding Documents**
*Project 2061 Benchmarks*
- *Objects can be described in terms of the materials they are made of (clay, cloth, paper, etc) and their physical properties (color, size, shape, weight, texture, flexibility, etc.).*
- *Rock is composed of different combinations of minerals. Smaller rocks come from breakage and weathering of bedrock and larger rocks. Soil is made partly from weathered rock, partly from plant remains and also contains many living organisms.*

*NRC Standards*
- *Soils have properties of color and texture, capacity to retain water, and ability to support the growth of many kinds of plants, including those in our food supply.*
- *Earth materials are solid rocks and soils, water, and the gases of the atmosphere. The varied materials have different physical and chemical properties, which make them useful in different ways, for example, as building materials, as sources of fuel, or for growing the plants we use as food. Earth materials provide many of the resources that humans use.*

**Science**
Earth science
soils
layers

**Integrated Processes**
Observing
Comparing and contrasting
Ordering

**Materials**
Scissors
Glue sticks
Student pages

**Background Information**
Soil scientists study soil profiles. These are the layers of the different types of soil. Driving through road cuts often reveal the various layers. The layers, which are called horizons, are identified with letters. The thickness of the horizons depends on the climate, topography, and plant life.

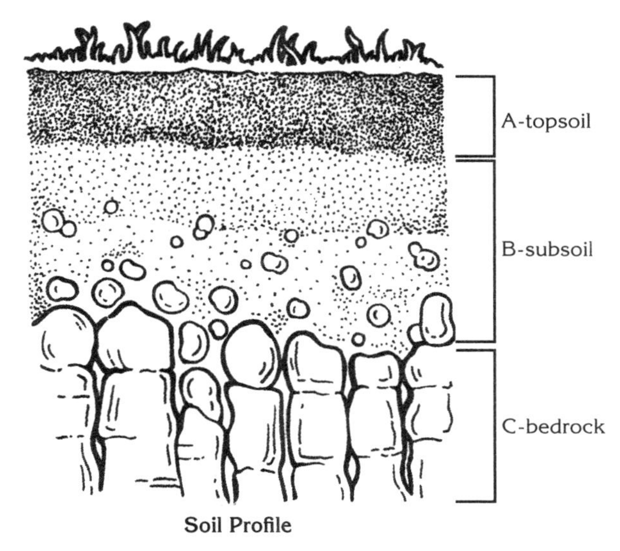

**Soil Profile**

A-horizon is the *topsoil* layer. It is rich in organic matter. The depth of this layer is usually not more than a meter or two. The soluble minerals from this layer leach down into the layer below it, leaving the insoluble minerals (e.g., quartz) and clays.

B-horizon, or the *subsoil*, is the transition zone between the A-horizon and the C-horizon. It contains the minerals that were leached out of the A-horizon. It has little organic matter. Some deep plant roots are found here.

The C-horizon is the decomposing *bedrock*, sometimes called parent material. Chemical weathering is largely responsible for this decomposition.

Soil scientists divide each of these horizons into sub-horizons. They also include other horizons such as O-horizon. This is a layer of organic material often found on top of the A-horizon. D-horizon is found below the C-horizon. It is unaltered bedrock. For this experience, students will investigate only three layers—the topsoil, subsoil, and bedrock layers.

 

## Management

1. Students can work together in small groups; however, each student needs to make his or her own record of layers.

## Procedure

1. Ask students if they have ever noticed the layers of soil along the highway as they ride through road cuts. Invite them to share their observations.
2. Ask the *Key Question* and state the *Learning Goals*. Tell them that they will be making a model that represents three different layers of soil like what can be seen in road cuts.
3. Inform students that they are going to read some information about different layers of soil and then use that information to put together a picture model of the different layers.
4. Distribute the student pages, scissors, and glue sticks. Go over the directions with the students.
5. After students have created their picture models, conclude with a discussion.

## Connecting Learning

1. How many layers are in our model? Name the layers.
2. Which layer is the oldest? [the bedrock] Where is it located in our model? [at the bottom]
3. The prefix *sub* means under or below. Why do you think the middle layer is called the subsoil? [It is located under the topsoil.]
4. When we plant flowers, in which layer do we plant them? [the topsoil]
5. The subsoil and bedrock make up our topsoil. What evidence can we find in the topsoil that shows that? [The topsoil has small pieces of rock that came from the bedrock and subsoil layers.]
6. How is our model like the real layers of the soil? [The model shows that there are different layers. It also shows the order of those layers.]
7. How is our model different than the real layers of the soil? [The real layers are different thicknesses. The real layers contain both living and nonliving things. Etc.]
8. Why would a farmer want to know about the layers of soil? [A farmer would want to have a thick layer of topsoil for growing healthy crops.]
9. What are you wondering now?

## Internet Connections

*Discovery Education: The Dirt on Soil*
http://school.discoveryeducation.com/schooladventures/soil/down_dirty.html
Has a picture of soil layers. The layers depicted as *weathered parent material* and *bedrock* are considered one layer in the activity.

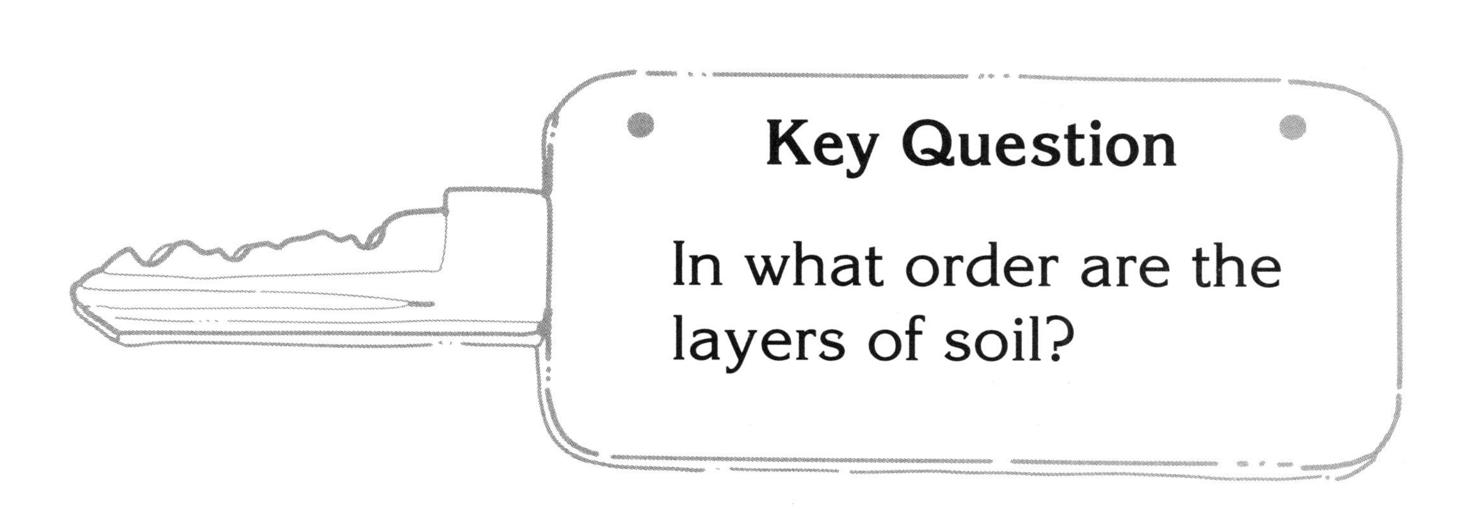


## Learning Goals

**Students will:**

- read descriptions of the three layers of soil, and
- make a soil model by cutting out and ordering the layers.

 

# A Record of Layers

Your soil profile will have three layers. Below are the descriptions of three layers. You need to decide the order of the pictures. You will build your model from the bottom layer up.

**Bottom Layer**
The bottom layer is the bedrock layer. It is made of rocks that are beginning to break up into smaller pieces. When you look at soil samples, you can see pieces of rock from which the soil is made. Find the picture that has the largest pieces of rock and glue it at the bottom of the model picture. Label it **bedrock.**

**Middle Layer**
The middle layer is called the subsoil layer. It is made up of soil and pieces of rock. There are a few deep plant roots that can be found in this layer. Cut out the picture that has smaller particles than the bedrock layer. Glue it to the top of the bedrock layer. Label this layer **subsoil.**

**Top Layer**
The top layer is called the topsoil. (That makes sense, doesn't it?) It is made of dirt, small rocks, and humus. There are many roots in this layer. There is also a lot of animal activity here. Earthworms and other animals use this layer to find food and shelter. Cut out the layer with the smallest particles. Glue it to the top of the subsoil. Label this layer **topsoil.**

Cut out the layers and put them in order on the next page.

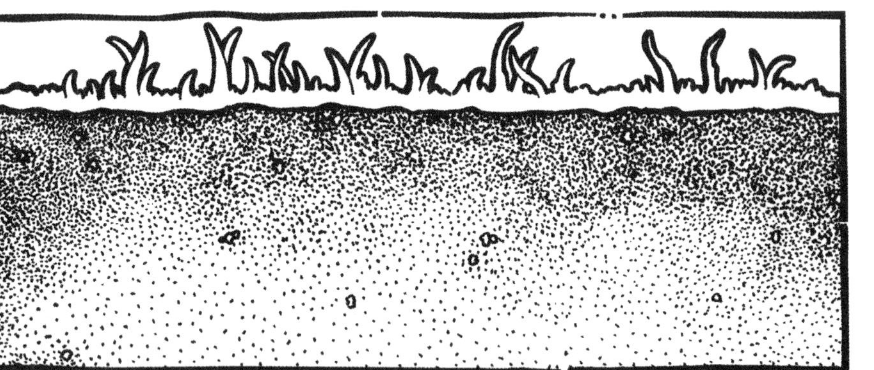

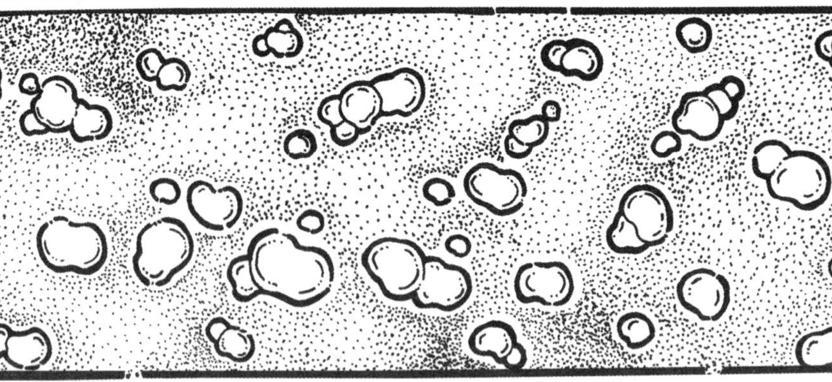

**Glue the layers in order.**

**Label the layers.**

CONNECTING LEARNING

## Connecting Learning

1. How many layers are in our model? Name the layers.
2. Which layer is the oldest? Where is it located in our model?
3. The prefix *sub* means under or below. Why do you think the middle layer is called the subsoil?
4. When we plant flowers, in which layer do we plant them?
5. The subsoil and bedrock make up our topsoil. What evidence can we find in the topsoil that shows that?

 

CONNECTING LEARNING

## Connecting Learning

6. How is our model like the real layers of the soil?
7. How is our model different than the real layers of the soil?
8. Why would a farmer want to know about the layers of soil?
9. What are you wondering now?

# Does This Hold Water?

**Topic**
Soil and water

**Key Question**
How well do different types of soil retain water?

**Learning Goals**
Students will:
- compare how different soils retain water,
- create and interpret graphs, and
- compare results of a scientific investigation.

**Guiding Documents**
*Project 2061 Benchmark*
- *Rock is composed of different combinations of minerals. Smaller rocks come from breakage and weathering of bedrock and larger rocks. Soil is made partly from weathered rock, partly from plant remains and also contains many living organisms.*

*NRC Standards*
- *Soils have properties of color and texture, capacity to retain water, and ability to support the growth of many kinds of plants, including those in our food supply.*
- *Earth materials are solid rocks and soils, water, and the gases of the atmosphere. The varied materials have different physical and chemical properties, which make them useful in different ways, for example, as building materials, as sources of fuel, or for growing the plants we use as food. Earth materials provide many of the resources that humans use.*

**Math**
Measurement
volume
time

**Science**
Earth science
soil

**Integrated Processes**
Observing
Comparing and contrasting
Communicating
Collecting and recording data

**Materials**
*For each group:*
3 Styrofoam cups, 6 oz
3 clear plastic cups, 10 oz
2 clear plastic cups, 9 oz
sand, potting soil, topsoil (see *Management 1)*
coffee filter, basket type
graduated strips (see *Management 2)*
colored pencils or crayons
scissors

**Background Information**
Soil texture is used to describe the proportions of different size particles in soil. The three particle sizes are sand, silt, and clay. Sand particles are the largest, silt particles fall in the middle, and clay particles are the smallest. The texture of soil is a very important property because it influences the soil's ability to hold water and air. Different particle sizes affect how well the soil drains. Sandy soils drain very rapidly as a result of the relatively large spaces between the particles. Soils with high clay content have very high water retention properties because there is relatively little space between the particles. It is very rare for soils to have only one type of particle size. We will use topsoil in this activity. It is a mixture of the sand, clay, and silt.

**Management**
1. Builders sand and topsoil can be purchased from landscape centers. Topsoil is going to vary depending on source of origin. It will be a mixture of sand, silt, and clay.
2. Copy the graduated strips onto transparency film. Cut out the strips and tape them to the appropriate sized clear plastic cups. If students have not read graduated scales such as these, take the time to practice by filling their cups to various levels and reading the scales. Notice that the spacings are not equal; this is a result of using a container with slanted sides.
3. Each group will need three graduated containers made from the 10-oz cups and two made from the 9-oz cups.
4. The soil samples should be wet prior to this investigation.

**Procedure**

1. Ask the *Key Question* and state the *Learning Goals.*
2. Distribute the three Styrofoam cups to each student group. Have them put five small holes in the bottom of each cup. A sharpened pencil can be used to make the holes.
3. Distribute the coffee filters. Direct the students to cup a circular portion from the filter to cover the holes in each cup.
4. Tell the students to place 100 mL of sand in the first cup, 100 mL of potting soil in the second cup, and 100 mL of topsoil in the third cup.
5. Place each Styrofoam cup inside a 10-oz plastic cup that has the graduated strip attached to its side.

6. Guide the students in a discussion on what makes a fair test. Inform them that they will be adding 100 mL of water to each type of soil. Distribute the 9-oz graduated cups. Make sure students understand how to read the scale. Question them about the intervals in the scale. [It's marked in 20 mL increments.] Ask students where they would have to fill the cup in order to have 50 mL of water. [about halfway between the line for 40 mL and 60 mL]
7. Tell them they will be timing to see when the first drops come through the holes in the bottom. Have students notice the clock dials on the student page. Ask them what they think the numbers represent. [seconds] Ask them how they think they would mark the clock dial if it took 20 seconds for the first drops to appear. [Make a mark at the top and a mark at 20. Use a line to connect each mark to the midpoint in the dial. Color in the wedge that they have drawn.]
8. Have each group test the three soils for how long it takes for the first drops to come out. Direct them to record the information on the student record sheet.
9. Direct the students to observe the three containers. Ask them what else they could record about the soil and water. (If no one suggests it, point out that they could also record how much water is in the bottom of the graduated cylinder).
10. Ask them how they could use subtraction to determine how much water was retained in each soil sample.
11. Direct the students to complete the graphs for each soil type.
12. Discuss any differences between the data the different groups collected.

**Connecting Learning**

1. How much time did it take for the first drop of water to drain through the soil? Was it the same amount of time for each cup? Explain.
2. Which type of soil seems to hold the most water? How could you justify this answer?
3. Who would want to know about how much water a certain type of soil is able to retain?
4. Why is it important to record observations in science?
5. How did the graphs help in this activity?
6. Did every group get the exact same results? Why is this important to talk about?
7. What are you wondering now?

 

# Does This Hold Water?

## Key Question

How well do different types of soil retain water?

## Learning Goals

**Students will:**

- compare how different soils retain water,
- create and interpret graphs, and
- compare results of a scientific investigation.

 

# Does This Hold Water?

**Time for the first drops to appear**

## Sand

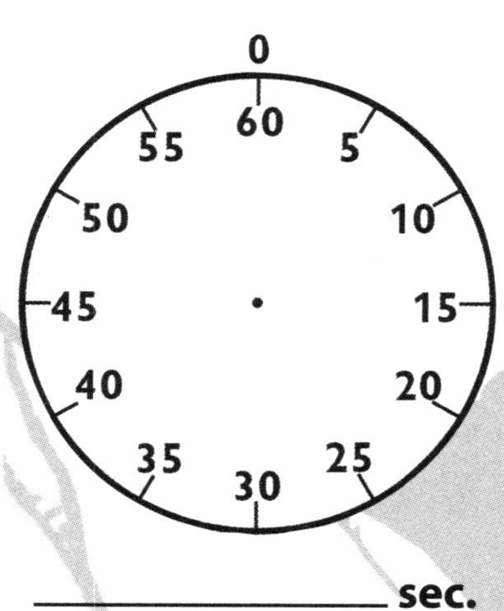


________________ sec.

## Potting Soil

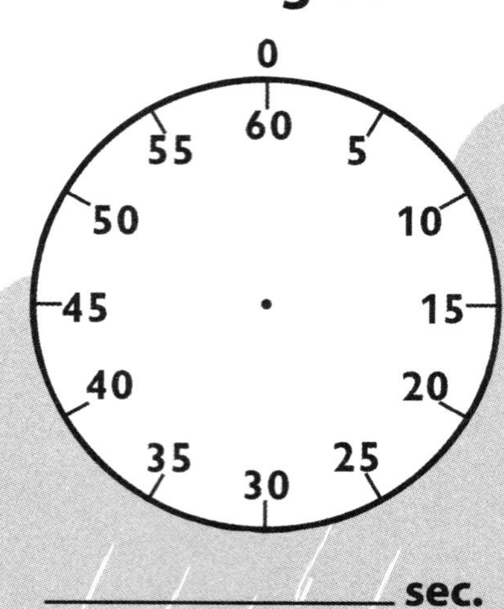


________________ sec.

## Topsoil

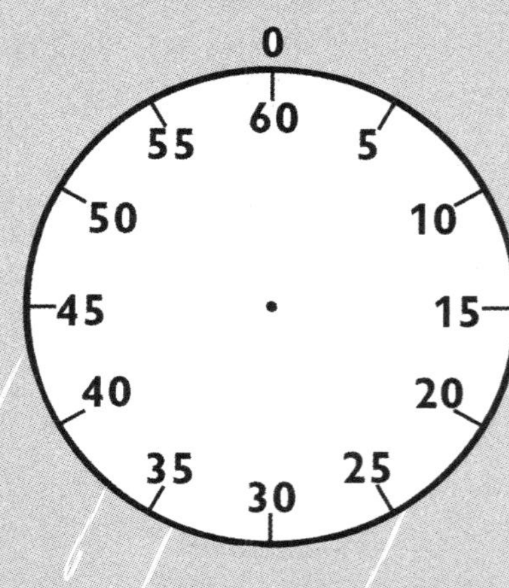


________________ sec.

## Water Strips

**Amount of water retained after 10 minutes**

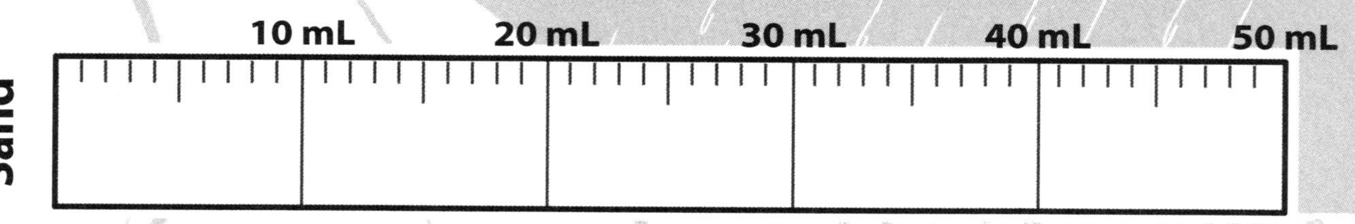


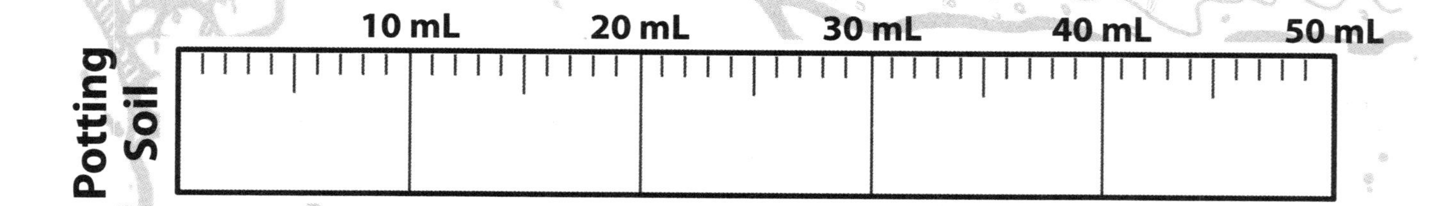


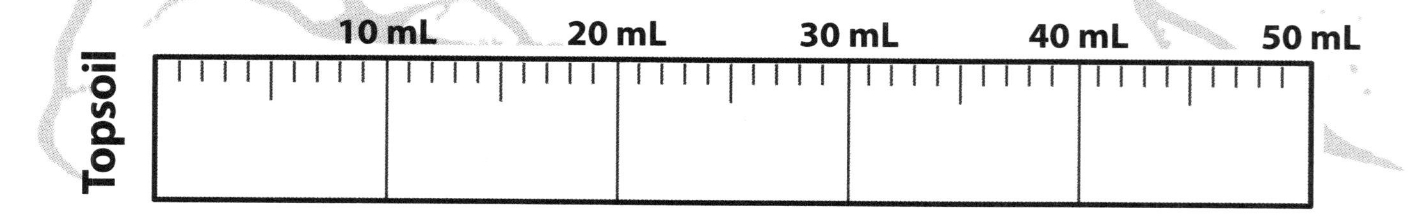


 

# Does This Hold Water?

**Cut the *Water Strips* from the first page and tape each one into a circle to create a circle graph. Record your data on this page.**

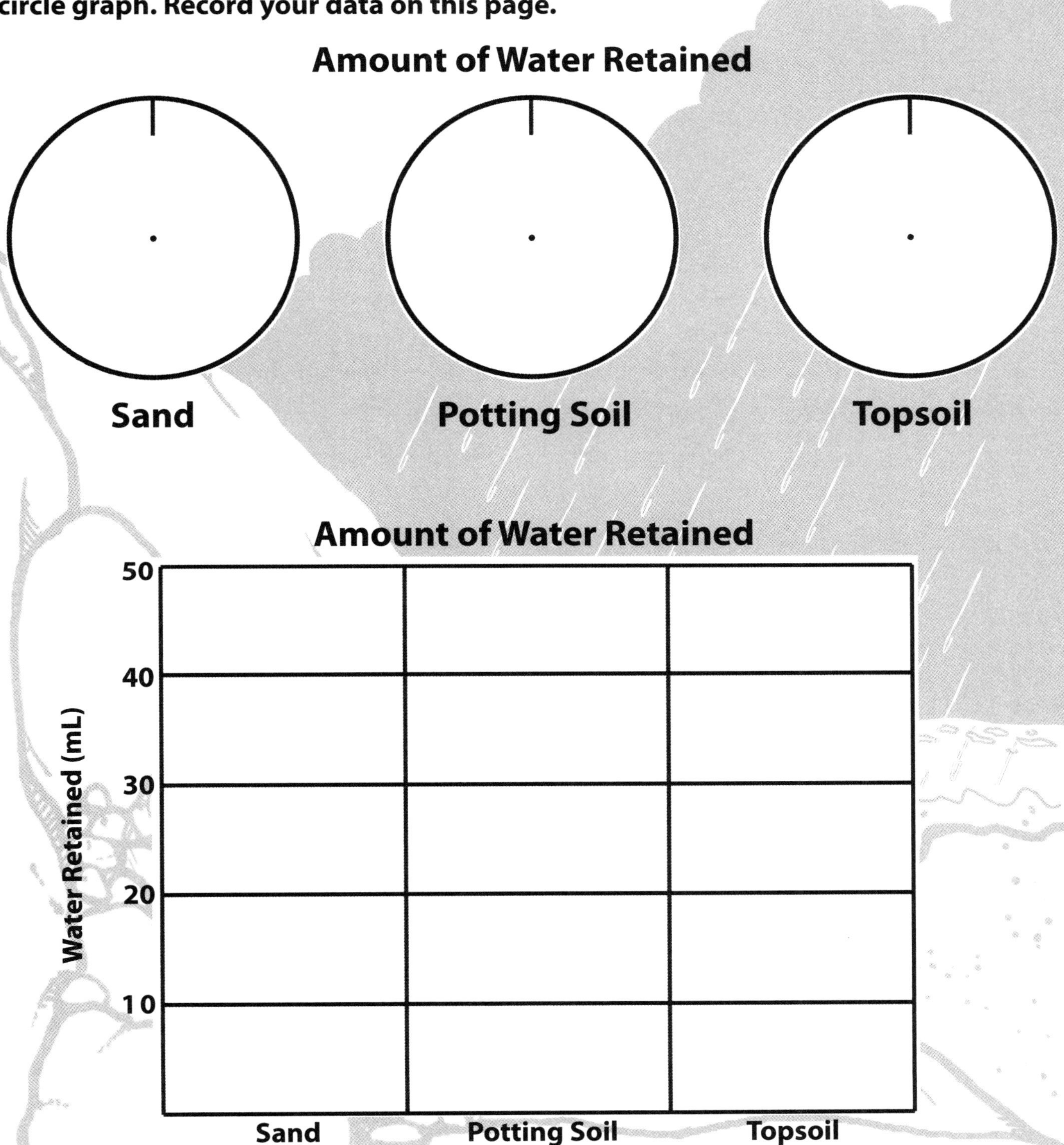


 

# Does This Hold Water?

**Copy these strips on overhead transparency film. Cut them apart and tape each strip to the appropriate sized plastic cup.**

| 240 | 240 | 240 | 240 | 240 | 240 | 240 | 240 | 240 | 240 |
|---|---|---|---|---|---|---|---|---|---|
| 220 | 220 | 220 | 220 | 220 | 220 | 220 | 220 | 220 | 220 |
| 200 | 200 | 200 | 200 | 200 | 200 | 200 | 200 | 200 | 200 |
| 180 | 180 | 180 | 180 | 180 | 180 | 180 | 180 | 180 | 180 |
| 160 | 160 | 160 | 160 | 160 | 160 | 160 | 160 | 160 | 160 |
| 140 | 140 | 140 | 140 | 140 | 140 | 140 | 140 | 140 | 140 |
| 120 | 120 | 120 | 120 | 120 | 120 | 120 | 120 | 120 | 120 |
| 100 | 100 | 100 | 100 | 100 | 100 | 100 | 100 | 100 | 100 |
| 80 | 80 | 80 | 80 | 80 | 80 | 80 | 80 | 80 | 80 |
| 60 | 60 | 60 | 60 | 60 | 60 | 60 | 60 | 60 | 60 |
| 40 | 40 | 40 | 40 | 40 | 40 | 40 | 40 | 40 | 40 |
| 20 | 20 | 20 | 20 | 20 | 20 | 20 | 20 | 20 | 20 |
| mL | mL | mL | mL | mL | mL | mL | mL | mL | mL |

9 oz

| 300 | 300 | 300 | 300 | 300 | 300 | 300 | 300 | 300 | 300 |
|---|---|---|---|---|---|---|---|---|---|
| 280 | 280 | 280 | 280 | 280 | 280 | 280 | 280 | 280 | 280 |
| 260 | 260 | 260 | 260 | 260 | 260 | 260 | 260 | 260 | 260 |
| 240 | 240 | 240 | 240 | 240 | 240 | 240 | 240 | 240 | 240 |
| 220 | 220 | 220 | 220 | 220 | 220 | 220 | 220 | 220 | 220 |
| 200 | 200 | 200 | 200 | 200 | 200 | 200 | 200 | 200 | 200 |
| 180 | 180 | 180 | 180 | 180 | 180 | 180 | 180 | 180 | 180 |
| 160 | 160 | 160 | 160 | 160 | 160 | 160 | 160 | 160 | 160 |
| 140 | 140 | 140 | 140 | 140 | 140 | 140 | 140 | 140 | 140 |
| 120 | 120 | 120 | 120 | 120 | 120 | 120 | 120 | 120 | 120 |
| 100 | 100 | 100 | 100 | 100 | 100 | 100 | 100 | 100 | 100 |
| 80 | 80 | 80 | 80 | 80 | 80 | 80 | 80 | 80 | 80 |
| 60 | 60 | 60 | 60 | 60 | 60 | 60 | 60 | 60 | 60 |
| 40 | 40 | 40 | 40 | 40 | 40 | 40 | 40 | 40 | 40 |
| 20 | 20 | 20 | 20 | 20 | 20 | 20 | 20 | 20 | 20 |
| mL | mL | mL | mL | mL | mL | mL | mL | mL | mL |

10 oz

 

# Does This Hold Water?

CONNECTING LEARNING

## Connecting Learning

1. How much time did it take for the first drop of water to drain through the soil? Was it the same amount of time for each cup? Explain.

2. Which type of soil seems to hold the most water? How could you justify this answer?

3. Who would want to know about how much water a certain type of soil is able to retain?

4. Why is it important to record observations in science?

5. How did the graphs help in this activity?

6. Did every group get the exact same results? Why is this important to talk about?

7. What are you wondering now?

**Topic**
Soil

**Key Question**
Which soil mixture grows the healthiest plants?

**Learning Goals**
Students will:
- raise plants from seeds in different soil mixtures, and
- evaluate which soil mixture promotes the best growth.

**Guiding Documents**
*Project 2061 Benchmark*
- *Rock is composed of different combinations of minerals. Smaller rocks come from breakage and weathering of bedrock and larger rocks. Soil is made partly from weathered rock, partly from plant remains and also contains many living organisms.*

*NRC Standards*
- *Soils have properties of color and texture, capacity to retain water, and ability to support the growth of many kinds of plants, including those in our food supply.*
- *Earth materials are solid rocks and soils, water, and the gases of the atmosphere. The varied materials have different physical and chemical properties, which make them useful in different ways, for example, as building materials, as sources of fuel, or for growing the plants we use as food. Earth materials provide many of the resources that humans use.*

**Math**
Measurement
 length

**Science**
Earth science
 soil
Life science
 plant growth

**Integrated Processes**
Observing
Comparing and contrasting
Controlling variables
Collecting and recording data
Interpreting data

**Materials**
Dirt/soil (see *Management 6)*
Sand
Peat moss
Planting containers (see *Management 4)*
Seeds
Zipper-type plastic bags, gallon size
Observation journals (see *Management 2)*
Potting mix recipe cards (see *Management 9)*
Watering cups
Centimeter rulers

**Background Information**
One of the properties of soil is its ability to support plant life. Plants obtain vital nutrients from the soil, and the quality of the soil directly affects how well plants will grow. Different kinds of plants have different soil needs, and a whole industry of specialized potting soils has been developed based on these needs. For cacti and other succulents, potting mixes have higher proportions of sand. African violet mixes contain perlite or vermiculite for improved soil aeration and drainage. Seed-starter potting mixes are mostly sphagnum peat moss, perlite, and fertilizer. This activity allows students to make their own potting soil mixes and see which are best suited for the plants they will be growing.

**Management**
1. This activity will take place over the course of several weeks. Plan to spend about an hour the first day creating the potting mixes and planting the seeds. Allow 10 minute every other day for watering and recording results. At the end of the activity, you will need to plan another block of time for discussion.
2. Each student will need an observation journal. The number of pages needed will depend upon how long you plan to have students observe their plants. Make one copy of the cover and questions page for each student and as many of the recording pages as necessary. Cut the pages in half, stack, and staple along the left side.
3. Select seeds that are easy to grow such as beans or peas. All students should use the same kind of seed. Soaking the seeds in water overnight will speed germination.
4. Each student needs his or her own planting container. Plastic or Styrofoam cups can be used as long as they have drainage holes poked in the bottom. Small three- or four-inch clay or plastic pots will also work. All students' containers need to be the same size.

 

5. Obtain enough garden soil, sand, and peat moss to make the soil mixes. The amount you will need depends on the sizes of the containers you are using and the number of students in your class.

| | Soil | Peat Moss | Sand |
|---|---|---|---|
| **Mix One** | 1 container | 1 container | 1 container |
| **Mix Two** | ½ container | ½ container | 2 containers |
| **Mix Three** | ½ container | 2 containers | ½ container |
| **Mix Four** | 2 containers | ½ container | ½ container |
| **Mix Five** | — | 2 containers | 1 container |
| **Mix Six** | — | 1 container | 2 containers |
| **Mix Seven** | 1 container | — | 2 containers |
| **Mix Eight** | 2 containers | — | 1 container |
| **Mix Nine** | 1 container | 2 containers | — |
| **Mix Ten** | 2 containers | 1 container | — |

6. Do not use potting soil. Collect soil from a flowerbed or garden. Try to find soil that is fairly uniform and free from rocks or large pieces of organic matter (twigs, dead leaves, insets, etc.).
7. Mark several clear plastic cups to be used as watering cups. In order to keep the variable of water constant, mark each cup at the same level to indicate the amount of water that is to be given to one plant.
8. You will need to identify a location (indoors or outdoors, if the weather permits) where students can place their plants to grow. It is important that the conditions be the same throughout this location—same temperature, same amount of sunlight, etc.
9. Copy the pages of recipe cards and cut them apart. You need one card for every three students. Recipes one and two are printed twice in case you have more than 30 students in your class. If you have 30 or fewer students, you will not need these extra cards.

**Procedure**

*Day One*

1. Divide the class into groups of three. There should be at least one group for each different potting mix you will be using. Give each group its recipe card and a plastic bag. Give each student a planting container. Instruct students to write their names and which potting mix they will be using on their containers.
2. Explain that each student is responsible for adding one ingredient of the soil mix to the bag. For the recipes that call for half a container, students will have to use their best judgment to fill a container halfway. If a recipe only has two ingredients, two students will share responsibility for adding the ingredient that calls for two containers. For example, in the group that has "Mix Six," one student will put a container full of peat moss in the bag. The remaining two students will each put one container of sand in the bag.
3. Once groups have placed the ingredients in the bags, instruct them to seal the bags and mix the parts thoroughly by shaking and turning the bags. After the parts are evenly distributed, have each student fill his or her planting container with the mix from the bag.
4. Distribute two seeds to each student, and have them plant the seeds on opposite sides of the containers. (This should ensure that at least one of the seeds will germinate).
5. Discuss the things that plants need to grow—soil, sunlight, water, nutrients, and space. Emphasize the importance of controlling all of these variables so that you can see what effect the different soils have on the plants. Ask students for their ideas on ways to do this.
6. Have students take their plants to the designated growing location. Be sure to put down plastic, newspaper, or something else to catch any water that may flow out of the bottoms of the containers if this area is not outside.
7. Distribute the watering cups and have students take turns carefully measuring and pouring the correct amount of water over their seeds.
8. Distribute the recording journals and have students record the day's date and their observations in the first space.

*Days Two and On*

1. Have students water their seeds as needed. If any students are absent, be sure their plants are also watered along with everyone else's.
2. If more than one of the seeds germinates, have students cut or pinch off the smaller of the plants at the ground level. Do not allow them to pull it out, as this could damage the roots of the other sprout. Removing the second plant reduces the competition for space.
3. Have students record their observations in their journals every other day. These observations should include soil condition, plant color, direction of plant growth, number and size of leaves, etc. Once their seeds sprout, have them measure and record the heights of their plants.
4. After two or three weeks of plant growth, discuss the results and see if a "best" mixture can be determined.

 

**Connecting Learning**

1. How long did the seeds take to sprout? Did the kind of soil seem to make a difference? Why or why not?
2. Which plants grew the quickest? Did the kind of soil seem to make a difference? Why or why not?
3. Which plants looked healthiest? Which soil mixes were those plants grown in?
4. Which plants looked the least healthy? Which soil mixes were those plants grown in?
5. What does this tell you about the kind of soil that our plants liked best?
6. Compare a healthy plant and one that is not as healthy. What could you do to make the unhealthy plant grow better? [add things to the soil, use fertilizer, etc.]
7. What variables did we need to control in our experiment? [kind of seed planted, container size, amount of water, container location, amount of sunlight] Why was it important to control these? [By controlling these variables, it is possible to see the effect that the different soils have on plant growth.]
8. What are you wondering now?

**Extensions**

1. Obtain vermiculite or perlite and add those to the soil mixes to see how this changes their quality and ability to support plant growth.
2. Add plant food to the plants in the poorest soils and see if it makes a difference.
3. Try a variety of different seeds to see if there are differences among plants in terms of what soils they like best.
4. Look at the ingredients of different kinds of potting soils and compare what they contain to the soil mixes you made.

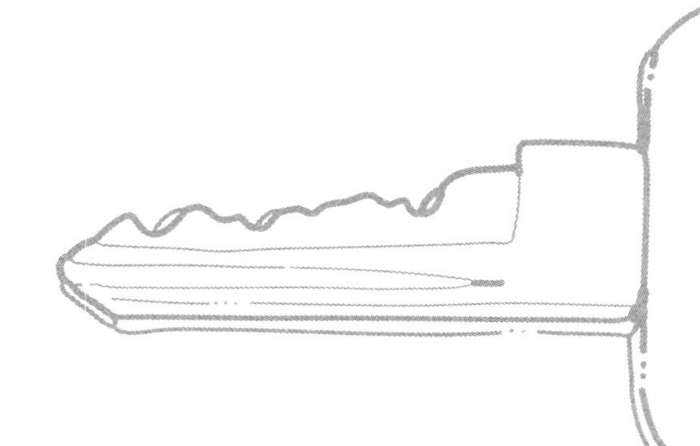

## Key Question

Which soil mixture grows the healthiest plants?

## Learning Goals

**Students will:**

- raise plants from seeds in different soil mixtures, and
- evaluate which soil mixture promotes the best growth.

# Don't MIX Me Up

*Ingredients:*
1 container soil
1 container peat moss
1 container sand

Put all the ingredients in the bag. Seal the bag and mix the ingredients. Fill your planting container with the mixture.

*Ingredients:*
½ container soil
½ container peat moss
2 containers sand

Put all the ingredients in the bag. Seal the bag and mix the ingredients. Fill your planting container with the mixture.

*Ingredients:*
½ container soil
2 containers peat moss
½ container sand

Put all the ingredients in the bag. Seal the bag and mix the ingredients. Fill your planting container with the mixture.

*Ingredients:*
2 containers soil
½ container peat moss
½ container sand

Put all the ingredients in the bag. Seal the bag and mix the ingredients. Fill your planting container with the mixture.

*Ingredients:*
2 containers peat moss
1 container sand

Put all the ingredients in the bag. Seal the bag and mix the ingredients. Fill your planting container with the mixture.

*Ingredients:*
1 container peat moss
2 containers sand

Put all the ingredients in the bag. Seal the bag and mix the ingredients. Fill your planting container with the mixture.

 

# Don't MIX Me Up

## Soil MIX 7

*Ingredients:*
1 container soil
2 containers sand

Put all the ingredients in the bag. Seal the bag and mix the ingredients. Fill your planting container with the mixture.

## Soil MIX 8

*Ingredients:*
2 containers soil
1 container sand

Put all the ingredients in the bag. Seal the bag and mix the ingredients. Fill your planting container with the mixture.

*Ingredients:*
1 container soil
2 containers peat moss

Put all the ingredients in the bag. Seal the bag and mix the ingredients. Fill your planting container with the mixture.

10

*Ingredients:*
2 containers soil
1 container peat moss

Put all the ingredients in the bag. Seal the bag and mix the ingredients. Fill your planting container with the mixture.

*Ingredients:*
1 container soil
1 container peat moss
1 container sand

Put all the ingredients in the bag. Seal the bag and mix the ingredients. Fill your planting container with the mixture.

2

*Ingredients:*
½ container soil
½ container peat moss
2 containers sand

Put all the ingredients in the bag. Seal the bag and mix the ingredients. Fill your planting container with the mixture.

1. When did your seeds sprout? Did they both sprout? How many days was this from when they were planted?
2. How much did your plant grow in the first week? How much did it grow in the second week?
3. Do you think your potting mix was good for your plant? Why or why not?
Don't MIX Me Up
Name
My soil mix

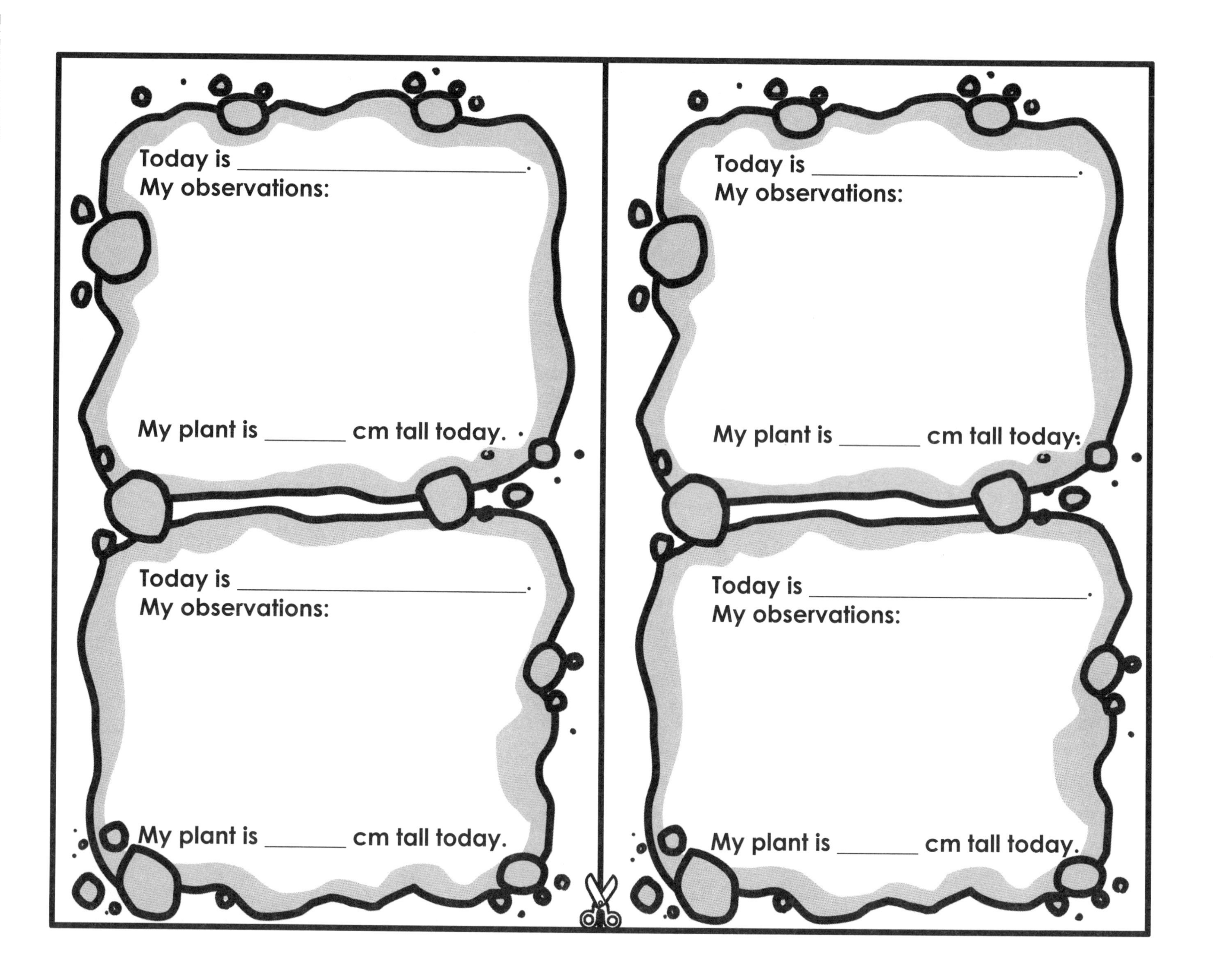
Today is ___.
My observations:
My plant is ___ cm tall today.
Today is ___.
My observations:
My plant is ___ cm tall today.
Today is ___.
My observations:
My plant is ___ cm tall today:
Today is ___.
My observations:
My plant is ___ cm tall today.

CONNECTING LEARNING

## Connecting Learning

1. How long did the seeds take to sprout? Did the kind of soil seem to make a difference? Why or why not?
2. Which plants grew the quickest? Did the kind of soil seem to make a difference? Why or why not?
3. Which plants looked healthiest? Which soil mixes were those plants grown in?
4. Which plants looked the least healthy? Which soil mixes were those plants grown in?

## Connecting Learning

5. What does this tell you about the kind of soil that our plants liked best?

6. Compare a healthy plant and one that is not as healthy. What could you do to make the unhealthy plant grow better?

7. What variables did we need to control in our experiment? Why was it important to control these?

8. What are you wondering now?

1

The Earth is often called the water planet. That is probably because water covers 70 percent of the Earth. The portion of the Earth's surface that is water is called the hydrosphere.

2

7

Water is one thing that makes our Earth special!

8

 

Water is all around you. It is in the air above you. It is in the ground below you.

3

The ocean contains most of the water on our planet. The water in our oceans is salt water.

4

We have fresh water in our shallow flowing rivers, deep lakes, green ponds, and clear streams.

5

We have frozen water in the form of glaciers. Only a very small amount of the Earth's water is fresh water. Most of it is frozen in ice caps.

6

Earth, my beautiful planet, what can I say?
I love your blue color, what makes you that way?

Is it a cloak that you've draped on your shoulders?
Or is it a blue dye spilled on your boulders?

Neither is true; it is the water that gives
The color of blue to this orb where we live.

Salt water, fresh water, frozen water too;
They all make our planet from space to be blue.

Each day we use water, why don't we run out?
My life depends on it, without any doubt.

The water it cycles, this much it is true.
There's no drop of water on Earth that is new.

Water that was here in days of Columbus
Is still here today to be shared among us.

There is as much water here today as before.
The reason for this is what you will explore.

# Checking on the Water Cycle

**Topic**
Water cycle

**Key Question**
What are the processes of the water cycle?

**Learning Goals**
Students will:
- build and observe a closed water cycle system, and
- identify the processes of the water cycle.

**Guiding Documents**
*Project 2061 Benchmark*
- *When liquid water disappears, it turns into a gas (vapor) in the air and can reappear as a liquid when cooled, or as a solid if cooled below the freezing point of water. Clouds and fog are made of tiny droplets of water.*

*NRC Standards*
- *Water, which covers the majority of the earth's surface, circulates through the crust, oceans, and atmosphere in what is known as the "water cycle." Water evaporates from the earth's surface, rises and cools as it moves to higher elevations, condenses as rain or snow, and falls to the surface where it collects in lakes, oceans, soil, and in rocks underground.*
- *Materials can exist in different states—solid, liquid, and gas. Some common materials, such as water, can be changed from one state to another by heating or cooling.*

**Science**
Earth science
  weather
    water cycle

**Integrated Processes**
Observing
Recording data
Comparing and contrasting
Relating
Inferring

**Materials**
*For each group of four:*
  2-L soda bottle, empty and clean
  clear plastic cup, 9 oz
  hot water in 5-oz Styrofoam cup
  ice
  scissors
  permanent markers, optional
  food coloring, optional

**Background Information**
A soda bottle containing water models the water cycle. Warmth causes the motion of water molecules to quicken; some escape their bonds, *evaporating* into the air and becoming invisible vapor. As the water vapor cools when it hits the cup of ice, the molecules slow down and *condense* again into a visible liquid. Water droplets may form at the top and sides and slowly drip or slide *(precipitate)* down, *accumulating* at the bottom of the bottle.

**Management**
1. Students can work in groups of four.
2. Hot tap water can be used. Adding food coloring to the water helps students see the accumulated water. It is also a teachable moment to let them discover that the water that condenses on the bottom of the ice-filled cup is clear, not colored.
3. If hot water is not available, coffee from the cafeteria or teacher's lounge can be used. Caution students to use care when working with the hot liquids.

**Procedure**
1. Hold a class discussion about where the rain comes from, how it gets in the clouds, etc. Lead students to thinking about the natural water cycle. Talk about the processes of evaporation, condensation, precipitation, and accumulation.
2. Tell students they will create a very simple water cycle in a soda bottle and look for evidence of the four processes.
3. Distribute a bottle and cup to each student or group. Have students cut approximately 5 centimeters (2 inches) off the top of the bottle. The resulting hole should be large enough that the 9-oz cup will sit in the top of the bottle, but not large enough that it will fall through.

 

4. Have students get the hot liquid in the 5-oz Styrofoam cup and pour it into the bottom of the soda bottle.
5. Direct them to fill the 9-oz plastic cup about half full of ice and set it in the top of the soda bottle.
6. Invite students to record the time at which their water cycle system was set up. Have them observe the system and record when they see moisture condense on the cup of ice. (This will happen very quickly.)
7. Have students continue to watch as the condensation drips off the cup (precipitates) and accumulates back in the bottom of the bottle.
8. Discuss how this model is like the natural water cycle and how it is different.
9. If desired, have students draw a scene on the outside of the bottle that represents the actual water cycle. They will need to use permanent pens.

**Connecting Learning**

1. What do you think will happen to the water in the bottle?
2. How could you tell that water had evaporated? [The condensed water droplets must have come from invisible water vapor in the air.]
3. Can you see evaporation? Explain. [No, water vapor is an invisible gas.]
4. What are the processes involved in the water cycle? [evaporation, condensation, precipitation, accumulation]
5. How does the mini water cycle demonstrate these processes? [evaporation—less water in the bottom of the bottle, condensation—water droplets on the cup, precipitation—water trickling down the sides, accumulation—water collecting at the bottom of the bottle]
6. What would happen if you left your mini water cycle in a warm place for one month? [The water will repeatedly cycle through the processes, but at a slower rate. Some of the water that evaporates may escape from the system if the cup isn't tightly fitted in the bottle.]
7. What are you wondering now?

**Curriculum Correlation**

*Literature*

Cole, Joanna. *The Magic School Bus Wet All Over: A Book About the Water Cycle.* Scholastic, Inc. New York. 1996.
Ms. Frizzle's class turns into raindrops. Join them as they evaporate, condense, rain, and make their way back to the ocean... only to evaporate all over again!

Locker, Thomas. *Water Dance.* Harcourt Brace & Co. Orlando. 1997.
Water speaks of its existence in such forms as storm clouds, mist, rainbows, and rivers. Includes factual information on the water cycle.

McKinney, Barbara Shaw. *A Drop Around the World.* Dawn Publications. Nevada City. 1998.
A raindrop cycles through liquid, solid, and vapor forms as it travels around the world.

Waldman, Neil. *The Snowflake: A Water Cycle Story.* The Millbrook Press. Brookfield, Connecticut. 2003.
Follow a snowflake through the changing phases of the water cycle—frozen pond, underground stream, irrigation system, cloud, ocean, etc.

 

# Checking on the Water Cycle

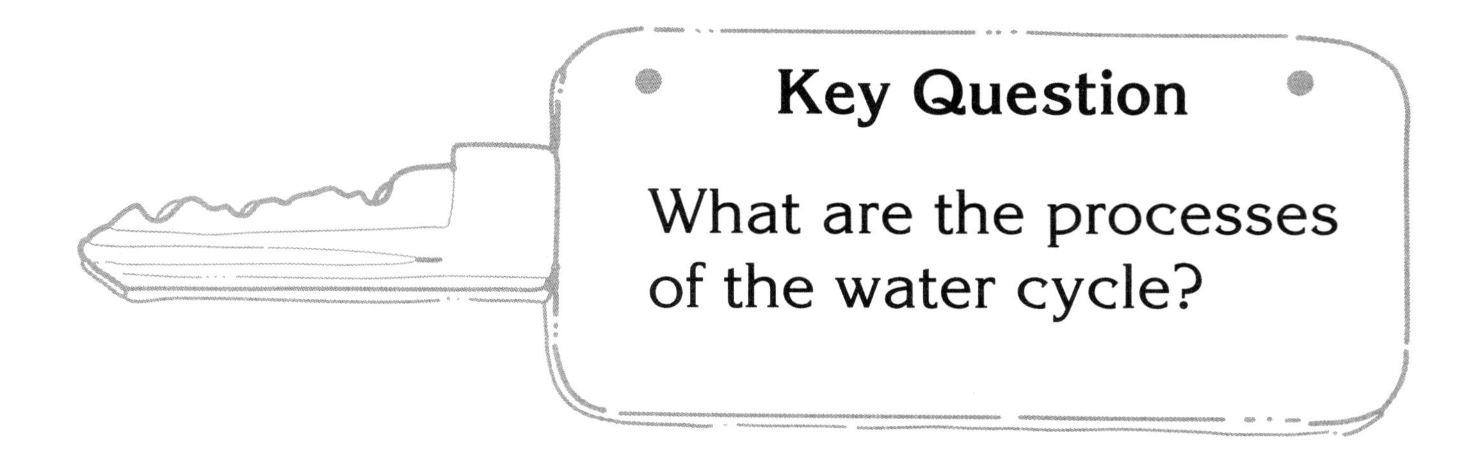


## Learning Goals

**Students will:**

- build and observe a closed water cycle system, and
- identify the processes of the water cycle.

 

# Checking on the Water Cycle

1. Use your scissors to cut 5 centimeters off the top of your bottle.
2. Check to see that a 9-oz cup will sit in the hole. If it doesn't, you may need to cut some more from the top of your bottle. Be careful; you don't want the cup to fall into the bottle.
3. Add hot liquid to the bottom of your bottle.
4. Fill your 9-ounce cup half full of ice and set it in the hole at the top of the bottle.
5. Write down the time at which your water cycle bottle was set up.
6. Watch carefully. Write down the time when you see moisture collect on the bottom of the cup. How long did this take?
7. When did you see drops of water fall back to the bottom of the bottle?
8. Illustrate what happened on the picture. Label the processes of evaporation, condensation, precipitation, and accumulation.

 

# Checking on the Water Cycle

CONNECTING LEARNING

## Connecting Learning

1. What do you think will happen to the water in the bottle?
2. How could you tell that water had evaporated?
3. Can you see evaporation? Explain.
4. What are the processes involved in the water cycle?
5. How does the mini water cycle demonstrate these processes?
6. What would happen if you left your mini water cycle in a warm place for one month?
7. What are you wondering now?

Dream On!
I had a dream and in it I saw
Precipitation was falling, was falling on all.
Solid or liquid, I did not know.
How should I dress? Was it rain or snow?
Hail or sleet could be falling outdoors.
Oh but wait, there are options galore.
Help me out of this nightmare, please.
Write all the forms you know underneath.
Then search this page over to see what you can find;
Are there forms here that didn't come to your mind?
Forms of Precipitation and Condensation
Look for eight forms.

**Topic**
Watersheds

**Key Question**
What happens when water runs downhill?

**Learning Goal**
Students will recognize that as water flows downhill it carries materials from the surrounding areas with it.

**Guiding Documents**
*Project 2061 Benchmarks*
- *Thing on or near the earth are pulled toward it by the earth's gravity.*
- *Waves, wind, water, and ice shape and reshape the earth's land surface by eroding rock and soil in some areas and depositing them in other areas, sometimes in seasonal layers.*

*NRC Standard*
- *The surface of the earth changes. Some changes are due to slow processes, such as erosion and weathering, and some changes are due to rapid processes, such as landslides, volcanic eruptions, and earthquakes.*

**Science**
Earth science
  watersheds
Physical science
  gravity

**Integrated Processes**
Observing
Comparing and contrasting
Applying

**Materials**
Cookie sheet (*see Management 1)*
Sand (see *Management 1)*
Glitter
Container to catch water
Pitcher of water

**Background Information**
In order for students to understand watersheds, they need to first conceptualize that water flows downhill because of the pull of gravity. As it flows downhill, it is affected by the ground's materials (rocks and soil); the steepness of the slope; the materials it picks up (erosion), transports, and deposits. Water will continue to flow downhill until there is something to stop it—a pond or a lake or the ocean.

**Management**
1. The cookie sheet needs to be jellyroll style, with edges. Fill the cookie sheet with a thin layer of sand from the sand box. If your school does not have a sand box, collect sand from another location or use playground dirt. Sprinkle some glitter on the sand at one end of the cookie sheet. The glitter will be used to represent pollutants.
2. Make sure the sand is a bit dirty.
3. Find a container to catch the water. It needs to be large enough to hold the width of the cookie sheet. Another cookie sheet, a baking dish, or an aluminum roasting pan would all work.

**Procedure**
1. Ask the *Key Question* and state the *Learning Goal.*
2. Tell students that they are going to be observing a simple demonstration using water. They will be challenged to share their observations of water that the demonstration shows.
3. Show students the cookie sheet filled with sand. Put the cookie sheet on a table and prop up the glittered end with books to a height of about four inches. Put the other end of the cookie sheet in the container to catch water.

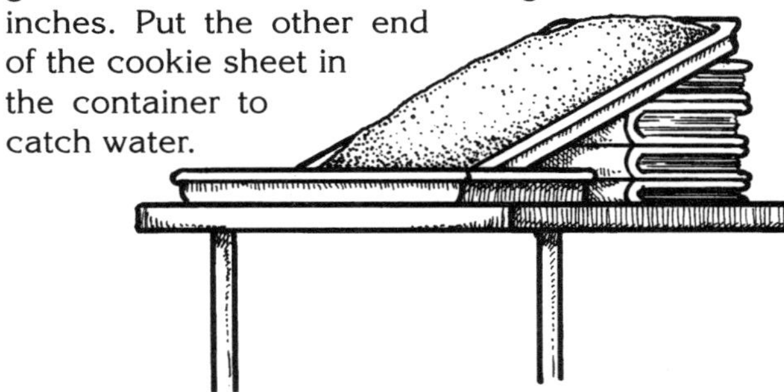

4. Have students observe the water in the pitcher. Ask them what they notice about the water. [It is clear, it is liquid, it takes the shape of the cup, etc.] Ask students what they think will happen when you pour the water on the sand in the cookie sheet. [The sand will get wet, the water will get dirty, it will go into the container at the bottom, etc.]
5. Slowly pour the water at the center of the top edge of the cookie sheet. When all of the water has flowed into the container at the bottom, have students examine it.

6. Ask students to make observations and describe what happened to the water, recording these on the board. Use questions to guide their thinking as necessary. [It flowed downhill through the sand and into the container at the bottom; it made the sand wet; it picked up some of the sand as it went; it started clear and ended dirty; it carried some of the glitter with it as it flowed downhill; etc.]
7. Revisit the list of observations that were generated.

**Connecting Learning**
1. What are some properties of water that you can observe?
2. What things did you observe during the demonstration?
3. How did the demonstration show that water flows downhill? Why does it flow downhill? [It is pulled downhill by gravity.]
4. What else did the demonstration show you about the water?
5. How is this demonstration like water in our real world?
   - What would the sand represent? [land]
   - What would the water in the pan represent? [ponds, lakes, oceans]
   - Where does the water come from? [rain, snow]
6. When it rains on hills, where does the water go? [Some soaks into the ground. Some may run down the hill.]
7. If the glitter represents pollution, what happens to the pollution that is upstream (on the high end of the model)? [It is transported downstream.] What are some substances that might be pollutants? [fertilizer from lawns or fields, oil from cars, spilled gasoline]
8. How do you think rivers form? How does this model help you understand that?
9. What are you wondering now?

**Extension**
1. Go outdoors after a rain to find real-world evidence of what the demonstration showed:
   - the ground is wet;
   - puddles take the shape of the holes they fill;
   - water travels downward (look for evidence of rivulets under trees or on slopes that aren't protected by grass).

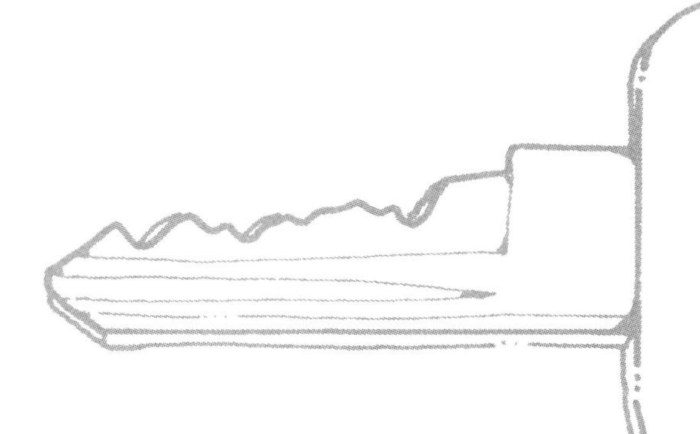

## Key Question

What happens when water runs downhill?

## Learning Goal

**Students will:**

recognize that as water flows downhill it carries materials from the surrounding areas with it.

# What's Seen Downstream?

Draw a scene that shows hills and water in a lake. Write a paragraph about what happens as water flows downhill.

# What's Seen Downstream?

Draw a scene that shows how water can change land. Write a paragraph about these changes.

## Connecting Learning

1. What are some properties of water that you can observe?

2. What things did you observe during the demonstration?

3. How did the demonstration show that water flows downhill? Why does it flow downhill?

4. What else did the demonstration show you about the water?

5. How is this demonstration like water in our real world?
   - What would the sand represent?
   - What would the water in the pan represent?
   - Where does the water come from?

CONNECTING LEARNING

## Connecting Learning

6. When it rains on hills, where does the water go?

7. If the glitter represents pollution, what happens to the pollution that is upstream (on the high end of the model)? What are some substances that might be pollutants?

8. How do you think rivers form? How does this model help you understand that?

9. What are you wondering now?

## Topic
Watershed

## Key Question
How can making a model help explain what a watershed is and how it works?

## Learning Goals
Students will:
- make a watershed model;
- identify a mountain, lake, river, and bay on the model; and
- spray the model with water and observe the flow of water over the model.

## Guiding Documents
*Project 2061 Benchmarks*
- *When liquid water disappears, it turns into a gas (vapor) in the air and can reappear as a liquid when cooled, or as a solid if cooled below the freezing point of water. Clouds and fog are made of tiny droplets of water.*
- *Waves, wind, water, and ice shape and reshape the earth's land surface by eroding rock and soil in some areas and depositing them in other areas, sometimes in seasonal layers.*

*NRC Standard*
- *The surface of the earth changes. Some changes are due to slow processes, such as erosion and weathering, and some changes are due to rapid processes, such as landslides, volcanic eruptions, and earthquakes.*

## Science
Earth science
 watershed
Physical science
 gravity

## Integrated Processes
Observing
Comparing and contrasting
Relating
Applying

## Materials
*For each group:*
 cardboard box lid
 plastic trash bag, kitchen size
 scissors
 transparent tape
 glue stick or transparent tape
 paper scraps
 plastic spray bottle
 paper towels

*For the teacher:*
 food coloring
 projection device

## Background Information
A *watershed* is the land that drains rainwater and melted snow to the same rivers, lakes, and bays. Water flows by gravity from higher to lower levels. Water also seeps into the ground. This replenishes the underground water that can be pumped to the surface for drinking, cooking, and washing.

Healthy watersheds provide clean water for recreational uses like canoeing or sailboating. Watersheds provide water for farmers to use to irrigate crops and for factories to manufacture products. Wildlife and plant reserves also depend upon healthy watersheds.

You can find your local watershed at http://www.water.epa.gov/type/watersheds/index.cfm

In this activity, students build a simple model from a box lid and a plastic trash bag. The teacher adds a drop of food coloring to the tallest mountain on the model, and the students spray the model with water and observe that the water flows off the mountains, into a river, and then into a lake or bay. Other areas on the model may collect water, which also model lakes.

## Key Vocabulary
*watershed:* an area of land that drains to the same rivers, lakes, bays, or other waterways
*river:* a flowing stream of water
*lake:* a large body of standing water
*bay:* a body of water partially enclosed by land but with a wide mouth, opening to the sea

 

**Management**

1. Organize the students into groups of two to four.
2. Collect cardboard box lids (the box lids from copy paper boxes are ideal for this activity). Each group will need one lid.
3. Each group will need one plastic trash bag. Cut each bag in half to make two large sheets of plastic for each group of students.
4. Each group will need a dozen sheets of scrap paper.
5. Either a glue stick or transparent tape can be used to construct the model.
6. Make a model ahead of time to show students the basic construction technique.

**Procedure**

*Part One: Making the Model*

1. Use the watershed picture and vocabulary page to review or introduce the watershed concept with the students (see *Background Information).*
2. Show and discuss your watershed model with the students.
3. Distribute a cardboard box lid, a plastic sheet, and a glue stick (or transparent tape) to each group.
4. Instruct the students to apply a line of glue to the top of one of the outer edges of the box lid and to press one edge of the plastic sheet along the line of glue.

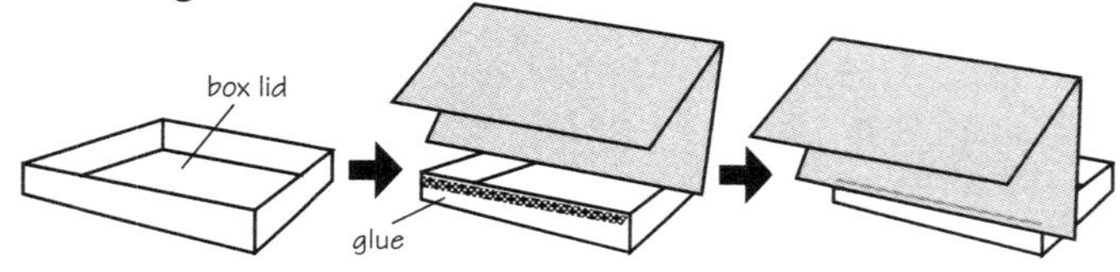


5. Direct the students to press the plastic sheet into the box so that the sheet fits neatly into the corners. Have them then glue the edges of the sheet to the outer sides of the box lid. This will waterproof the box lids.
6. Move from group to group and assist the students in lining the box lids.

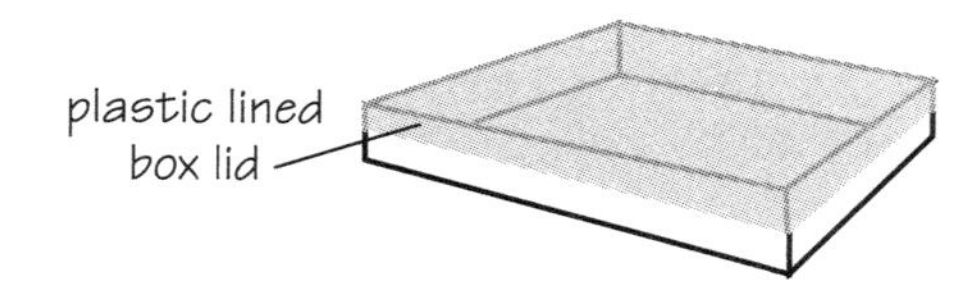


7. Distribute four sheets of scrap paper to each group. Tell the students to crumple each sheet into a ball and to build a mountain in one corner of the box lid and to glue the paper wads to each other and to the plastic liner.

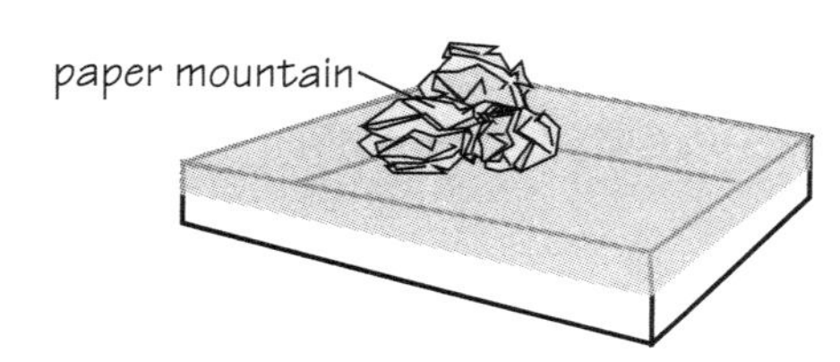


8. Distribute another sheet of paper to each group. Show them how to airplane-fold the paper to form a river channel. Tell them to glue the channel near the base of the mountain, pointing toward the opposite corner of the box lid.

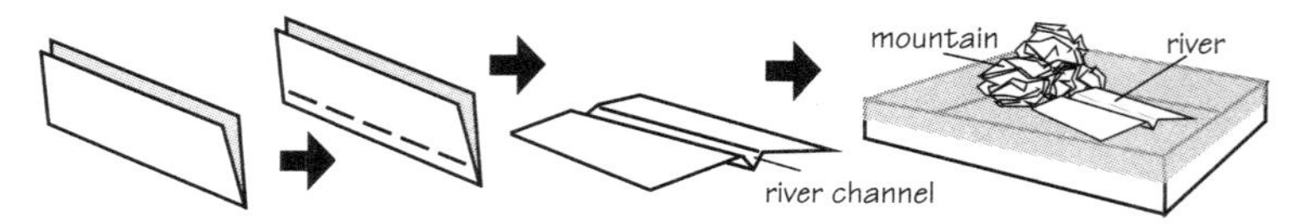


9. Distribute three more sheets of scrap paper to each group. Have them tear each sheet in half, make paper wads and flatten, and then glue the flattened paper wads from the mountain toward the opposite corner. Tell them not to cover any part of the river channel.

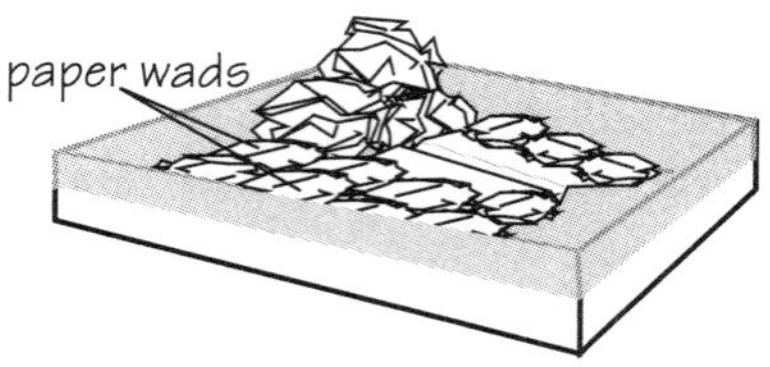


10. Distribute the second sheet of plastic to each group. Direct the students to glue one edge of the plastic sheet to the edge of the box lid behind the mountain. Tell them to apply glue to the top of the paper landforms in the bottom of the box, carefully fold over the plastic sheet and press down on the sheet so that it sticks to the glued areas.
11. Move from group to group and verify that each group has completed a waterproof model.

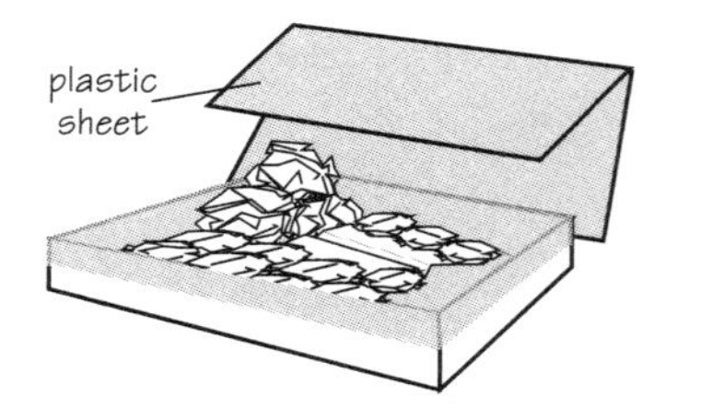


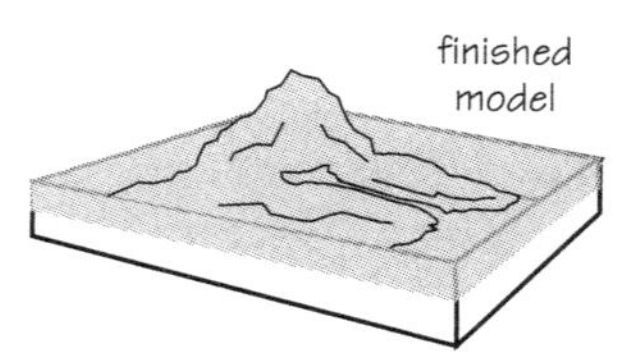


*Part Two: Using the Model*

1. Distribute the student page to every student.
2. Tell the students to draw the landform they created in the box on the diagram of the box.
3. Move from group to group and place a drop of food coloring in a notch on each mountain. (If needed, press a notch into the mountain.)
4. Distribute a spray bottle containing water to each group. Instruct one student in each group to spray water on the top of the model. Caution them to hold bottle at least six inches from the model. Tell the students that the spray bottle simulates rainfall.

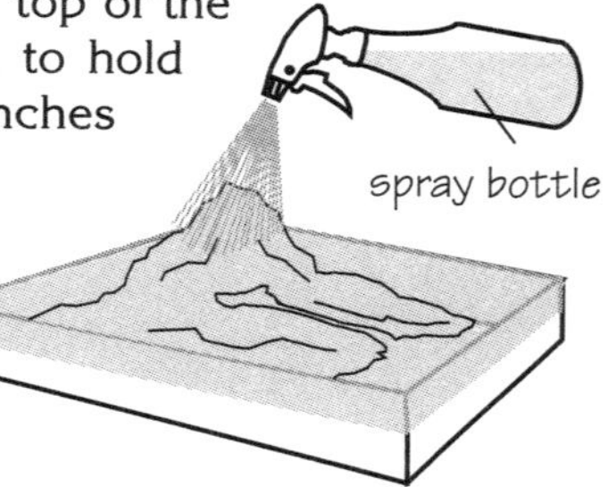


 

5. Have the students finish answering the questions on the page.

**Connecting Learning**

1. Name the features of your model. [mountain, rivers, lakes]
2. What happened when you sprayed water on your model? [The water flowed to the bottom of the model.]
3. What caused the water to flow in that direction? [gravity]
4. What is a river?
5. What is a watershed? [an area of land that drains to the same rivers, lakes, bays, or other waterways]
6. Was all of your model one watershed? Explain. (It will depend on the model. Some mountains may drain in different directions.)
7. What is meant by the statement "We all live downstream"? [Water flows from higher to lower levels. Unless you live at the highest point in a watershed, water is flowing to where you live. This means you live downstream.]
8. How is the watershed affected by what is upstream? [If pollution gets in the water up in the mountains, it will end up in the bay or ocean.]
9. What are you wondering now?

**Internet Connections**

*U.S. Environmental Protection Agency*
http://www.water.epa.gov/type/watersheds/index.cfm
All sorts of resources on watersheds. Lets you enter your zip code to find your local watershed.

*How Stuff Works*
http://www.maps.howstuffworks.com/united-states-watersheds-maps.htm
Contains a U.S. watershed map that can be viewed regionally.

## Key Question

How can making a model help explain what a watershed is and how it works?

## Learning Goals

**Students will:**

- make a watershed model;
- identify a mountain, lake, river, and bay on the model; and
- spray the model with water and observe the flow of water over the model.

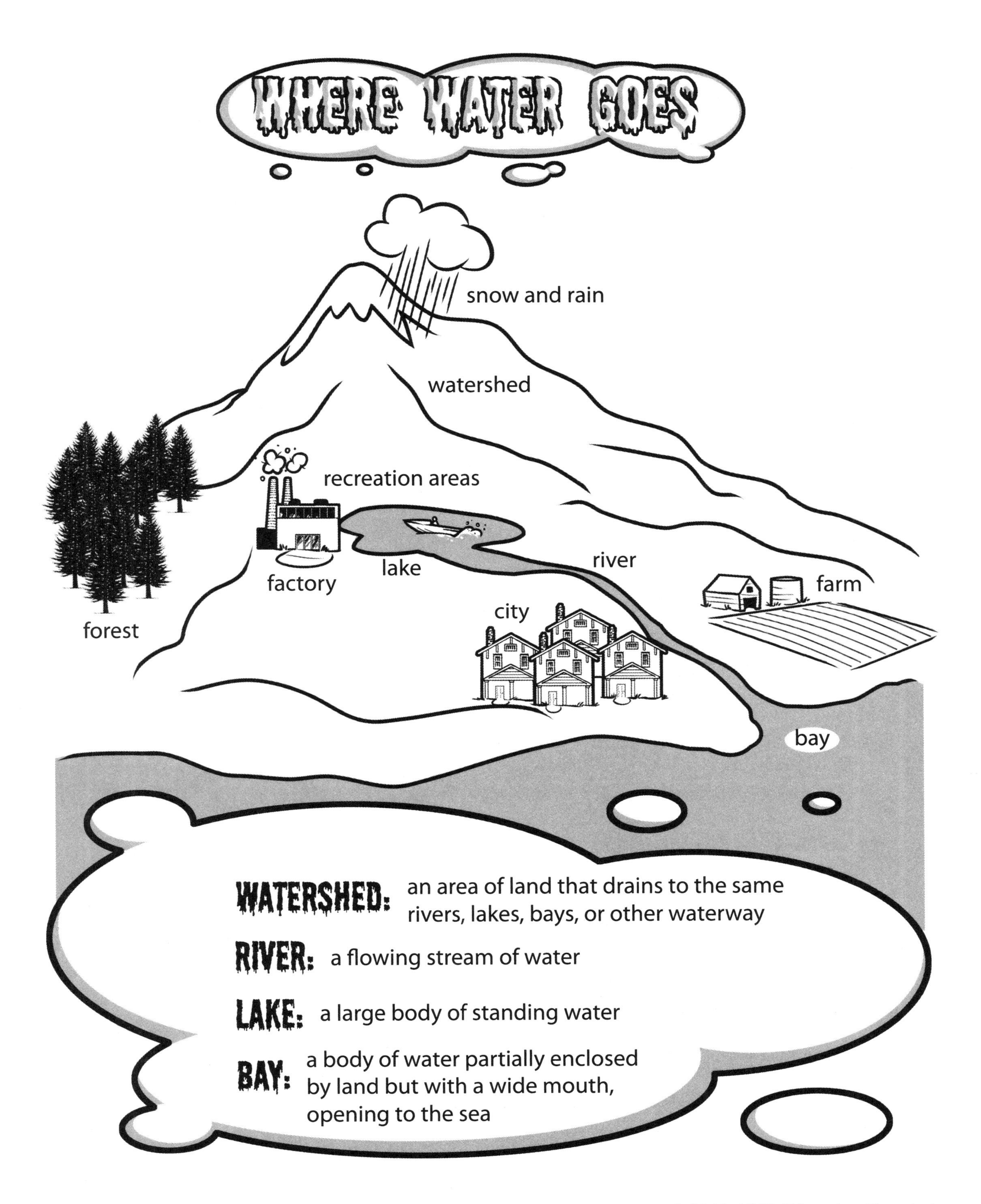
WHERE WATER GOES
snow and rain
watershed
recreation areas
river
lake
factory
farm
city
forest
bay
WATERSHED: an area of land that drains to the same rivers, lakes, bays, or other waterway
RIVER: a flowing stream of water
LAKE: a large body of standing water
BAY: a body of water partially enclosed by land but with a wide mouth, opening to the sea

 

# WHERE WATER GOES

1. Make a drawing in the cardboard box of your model.

2. Identify and label the highest mountain, a lake, a river, and a bay on your drawing.

3. After your teacher has placed a drop of food coloring on the tallest mountain of your model, spray the model with water. This models rain falling.

4. Describe what happened when you sprayed water over your model.

5. Color your drawing to show how rainfall moved on the model.

6. Describe how the "rainfall" flowed over your model.

7. Where did most of the rainfall go?

8. Explain what this statement means: "We all live downstream."

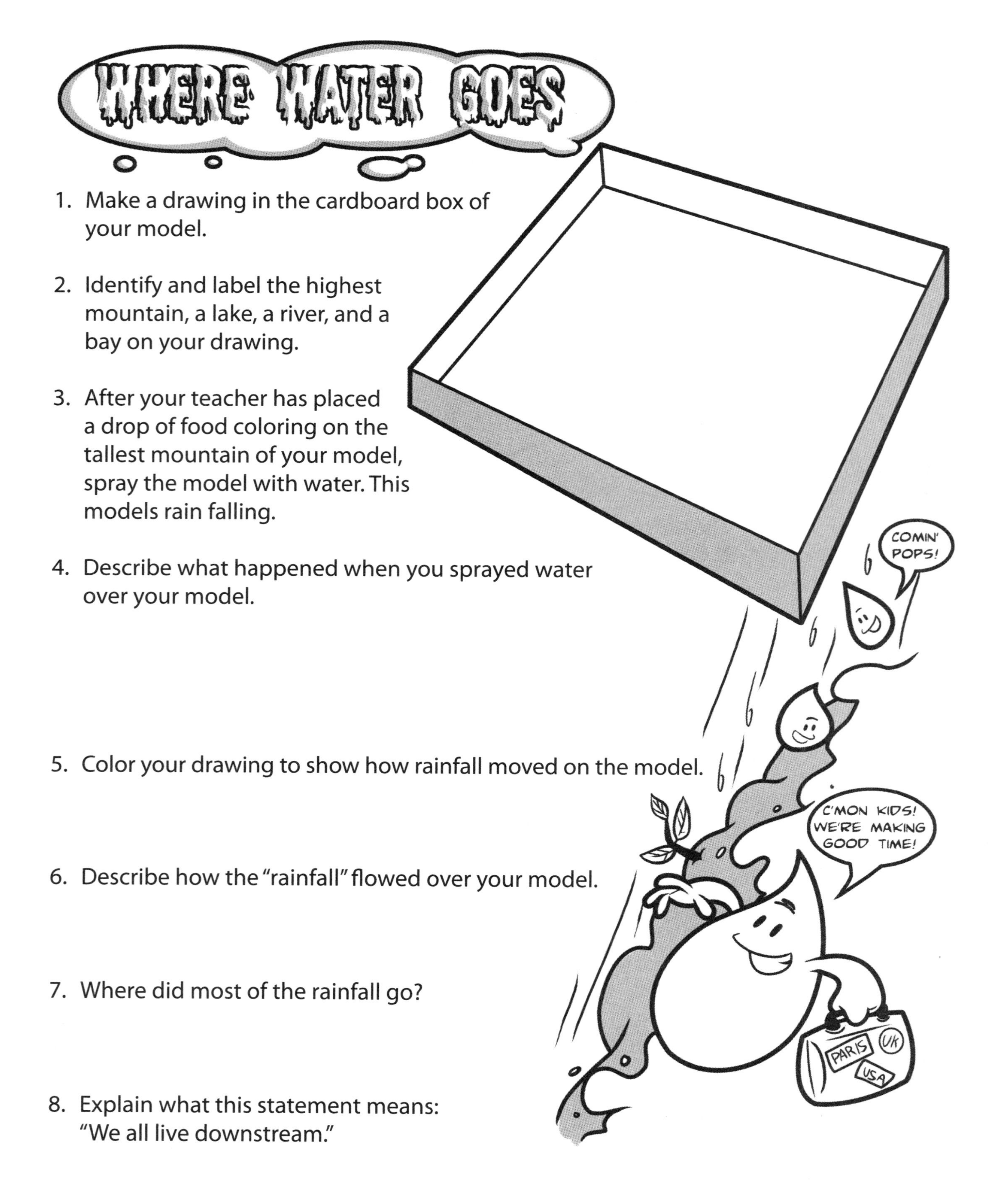

## Connecting Learning

1. Name the features of your model.
2. What happened when you sprayed water on your model?
3. What caused the water to flow in that direction?
4. What is a river?
5. What is a watershed?
6. Was all of your model one watershed? Explain.
7. What is meant by the statement “We all live downstream”?

## Connecting Learning

8. How is the watershed affected by what is upstream?

9. What are you wondering now?

 

# Changes, Fast and Slow

1

2

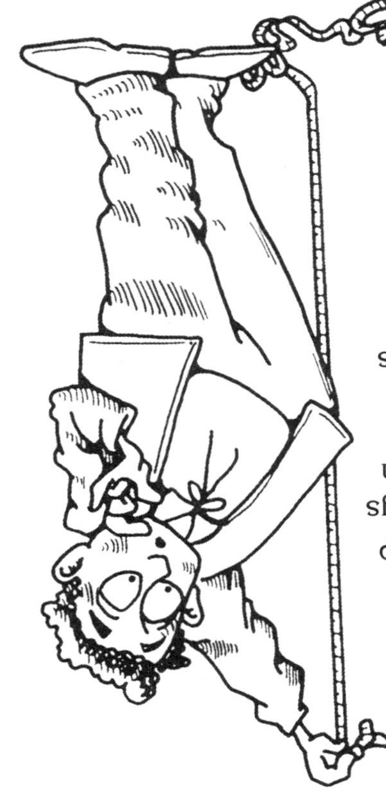

Are you the same height you were last year? Is the weather the same? You change. Things change. The Earth changes. Some changes are fast, and some changes are slow.

11

Think about a jar with very dirty water in it. If you shake that jar, the dirt stays mixed in the water. As soon as the water begins to slow down, however, the dirt settles to the bottom. The same thing happens to the river. The river picks up dirt and carries it. When the flow slows down, the river deposits its load. Deltas are created at the river's end.

Changes come
and changes go.
Some are fast, and
some are slow.
Earth is changing
at its own pace.
I'm just glad it's
not a race.

12

What evidence can you find of Earth's changes? The most dramatic changes are from the eruptions of volcanoes, from earthquakes, and from landslides. These are rapid processes because they can change an area in just a few seconds. They can be very sudden and very dangerous.

3

Then there are slow changes. Sometimes they are so slow that we don't even take notice of them. And because we don't notice them, they need to be pointed out. Weathering, erosion, and deposition are examples of slow changes.

4

Water is the most powerful method of erosion. Have you ever seen television reports of floods? They show cars and houses being picked up and carried off. Moving water cuts through soil and carries it off. It moves rocks and sand along beaches.

9

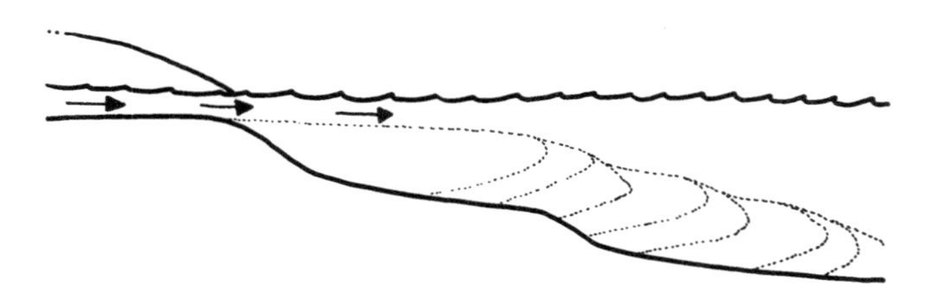

The materials the water picks up as it moves along are deposited in other places. In the case of rivers, the major load of materials is deposited at the place where the river enters a lake or ocean. This is where the river slows down. Why here?

10

Weathering is the breaking down of rocks. Rocks can be weathered in several ways. Water expands as it freezes. If it seeps into the cracks of a rock and freezes, it can break off that bit of rock. Plant roots can break rocks as they grow. Burrowing animals can break off bits of rocks. Rain that picks up chemicals from rocks and decaying organisms can cause chemical reactions that will break down rocks. While weathering may be an unwanted process on our stone buildings and monuments, it is a process that is necessary for making soil.

5

The movement of wind, ice, or water causes erosion. The movement of these things picks up soil and bits of rock and takes them somewhere else. Think about what happens on a windy day. The wind may pick your hat off your head and blow it across the playground, or it may pluck your homework from your arms and blow it down the street.

6

The wind does the same thing to soil and sand. It picks them up and transports them to other places. They are deposited (put down) in some place other than where they began. Look in different nooks and crannies around the outside of your school building. Do you see any piles of dirt or sand? These were probably picked up, carried, and deposited there by the wind.

7

Ice will do the same thing, but this isn't as easy for us to see. Large sheets of ice, called glaciers, are pulled down mountainsides by the force of gravity. As these large sheets of ice move, they pick up and carry rocks and soil with them.

8

# WEATHERING

This is a drawing of El Capitan, a huge rock face in Yosemite National Park in California.

1

One route to climb this rock face is 1900 feet long. If you were to use this route, the distance you would have to climb upward would equal 380 five-foot tall students standing on each other's heads to reach the top of that rock face. But did you know that this huge rock of granite will one day be weathered away into pebbles of sand?

2

Acid rain is an example of chemical weathering. The rain combines with carbon dioxide in the air and from decaying plants on the ground. This produces an acid that changes rock.

7

The results of weathering caused by acid rain are easily seen on buildings that are made of limestone or statues that are made of marble. The limestone and marble get very porous, in other words, they develop holes.

8

What does *weathered away* mean? That means that Earth's solid surface will gradually be broken into smaller and smaller pieces. It happens all the time; it's a natural action.

3

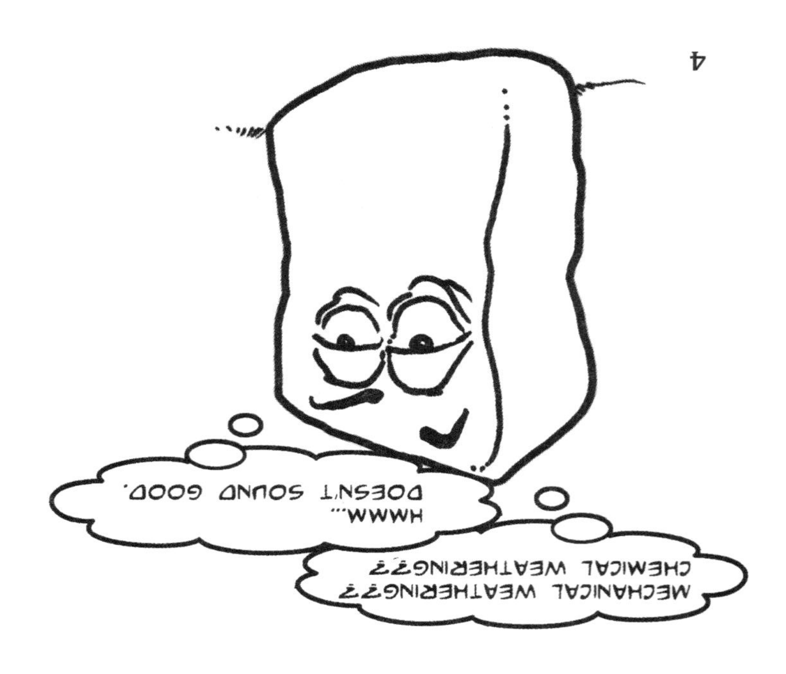


How does this happen? There are two main types of weathering. One is called mechanical weathering and the other is called chemical weathering.

4

Mechanical weathering occurs when physical forces such as water freezing, wind moving, and plant roots growing, break up rocks.

5

Chemical weathering changes the materials that make up the rocks. Water is the major agent of chemical weathering. Pure water will not change the rock's materials. But water picks up chemicals in the air and on the land. When mixed with water, some of these chemicals can change rocks.

6

**Topic**
Weathering

**Key Question**
What are some the ways the surface of the Earth can be physically changed?

**Learning Goal**
Students will model how sand erodes the surface of the Earth.

**Guiding Documents**
*Project 2061 Benchmark*
- *Change is something that happens to many things.*

*NRC Standard*
- *The surface of the earth changes. Some changes are due to slow processes, such as erosion and weathering, and some changes are due to rapid processes such as landslides, volcanic eruptions, and earthquakes.*

**Science**
Earth science
physical weathering

**Integrated Processes**
Observing
Comparing and contrasting
Applying

**Materials**
*For each student group:*
chalk (see *Management 1)*
sandpaper (see *Management 2)*
hand lens

**Background Information**
Mechanical weathering occurs whenever an external force is applied to a rock that causes the rock to break down. There are several different types of mechanical weathering. This activity will explore the physical process of abrasion—the action of small pieces of rock hitting the surfaces of larger rocks and breaking them down. The continual impact and friction from these small pieces will gradually grind away the larger rock. This process takes place both on the land surfaces of the Earth as well as in water. The sandpaper used in this activity will be used as a model to explore how small particles can affect larger Earth materials. The chalk used is the mineral calcite.

**Management**
1. White chalk works best for this activity.
2. Use a medium-grain (100 or 120) sandpaper. You may also want to have coarse- (60 or 180) and fine-grain (220 or 250) samples for the students to examine.

**Procedure**
1. Ask the *Key Question* and state the *Learning Goal.*
2. Distribute the materials to each student group. Ask the students to use the hand lens to examine the surface of the sandpaper as well as the piece of chalk.
3. Direct the students to scrape the surface of the chalk with the sandpaper. Tell them to observe the chalk and the piece of sandpaper.
4. Allow time for the students to respond to questions 4 through 7 on the student sheet.

**Connecting Learning**
1. What did the sandpaper and chalk represent in this activity? [sandpaper represents sand, chalk represents rocks]
2. What evidence do you have that the chalk was being changed?
3. How is this like blowing sand?
4. What would happen if you had sandpaper that had larger pieces of sand on it? What about smaller pieces?
5. What are you wondering now?

 

# Weathering Away

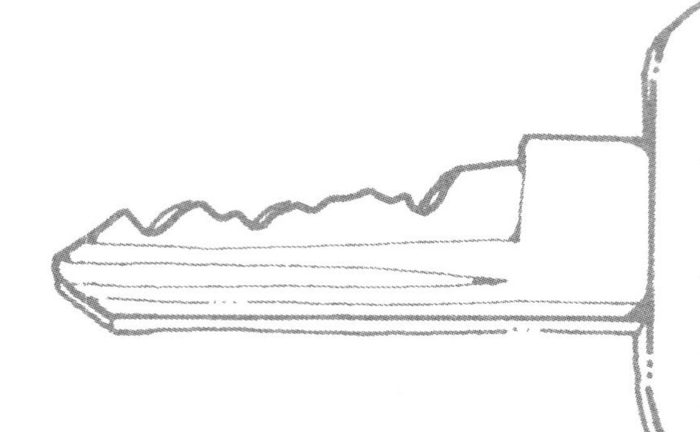

**Key Question**

What are some the ways the surface of the Earth can be physically changed?

## Learning Goal

**Students will:**

model how sand erodes the surface of the Earth.

 

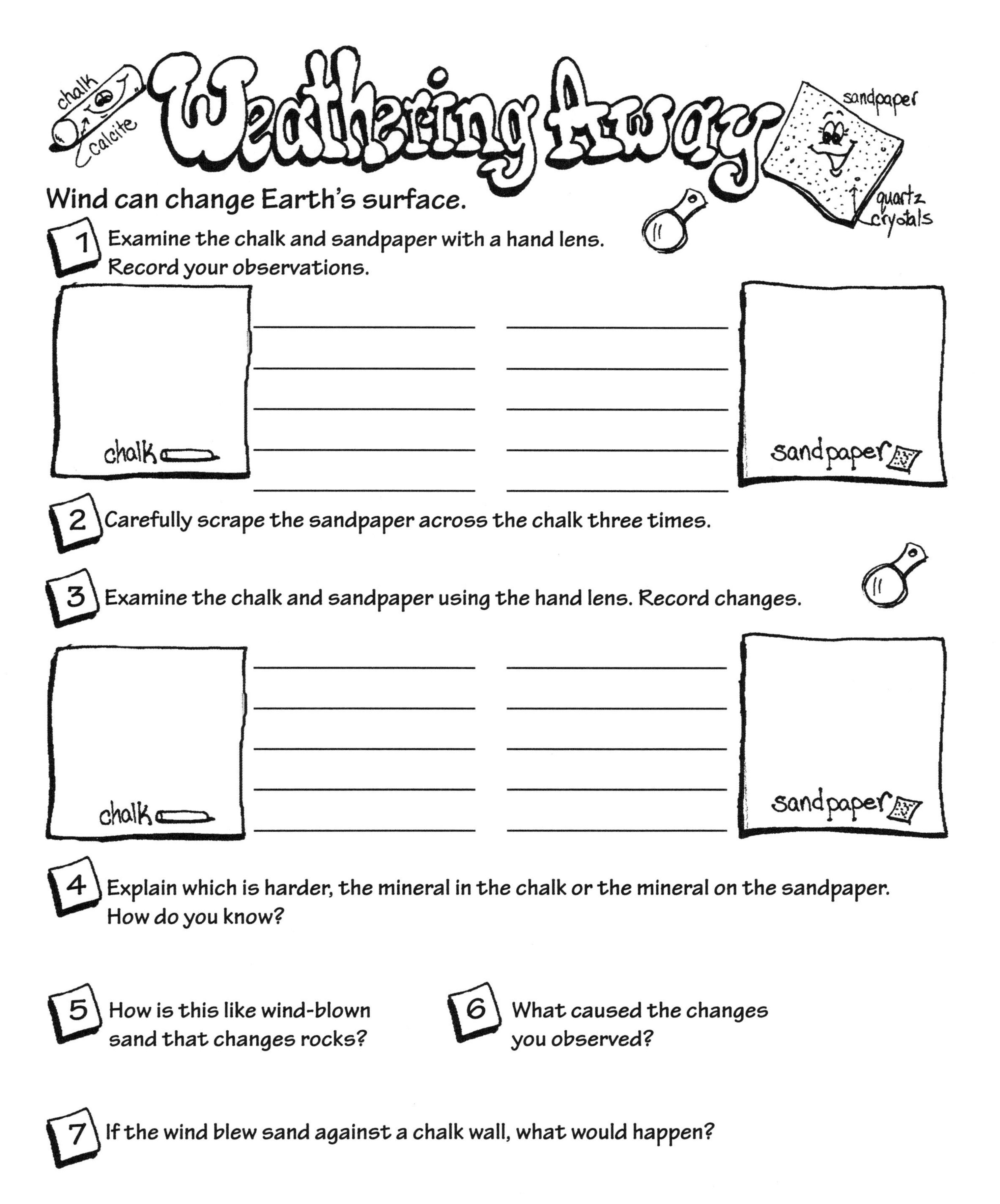


# Weathering Away

**Wind can change Earth's surface.**

1. Examine the chalk and sandpaper with a hand lens. Record your observations.

chalk

sandpaper

2. Carefully scrape the sandpaper across the chalk three times.

3. Examine the chalk and sandpaper using the hand lens. Record changes.

chalk

sandpaper

4. Explain which is harder, the mineral in the chalk or the mineral on the sandpaper. How do you know?

5. How is this like wind-blown sand that changes rocks?

6. What caused the changes you observed?

7. If the wind blew sand against a chalk wall, what would happen?

## Connecting Learning

1. What did the sandpaper and chalk represent in this activity?
2. What evidence do you have that the chalk (rock) was being changed?
3. How is this like blowing sand?
4. What would happen if the sandpaper had larger pieces of sand on it? What about smaller pieces?
5. What are you wondering now?

## Topic
Weathering

## Key Question
How does weathering change the surface of the Earth?

## Learning Goals
Students will:
- explain the causes of weathering, and
- create models depicting various types of weathering.

## Guiding Documents
*Project 2061 Benchmarks*
- *People can often learn about things around them by just observing those things carefully, but sometimes they can learn more by doing something to the things and noting what happens.*
- *Change is something that happens to many things.*
- *A model of something is different from the real thing but can be used to learn something about the real thing.*

*NRC Standard*
- *The surface of the earth changes. Some changes are due to slow processes, such as erosion and weathering, and some changes are due to rapid processes, such as landslides, volcanic eruptions, and earthquakes.*

## Science
Earth science
  weathering

## Integrated Processes
Observing
Collecting and recording data
Comparing and contrasting
Predicting
Applying

## Materials
Several small rocks (see *Management 3)*
One large plastic jar with lid (see *Management 4)*
Paper towels
Plastic bags, zipper type, pint size
Plastic eggs, one per group
Tub for water
Plastic cup
Ice cubes
Containers for water (see *Management 5)*
Source of hot water (see *Management 6)*
Access to a freezer (see *Management 7)*
*Weathering Ways* journals (see *Management 8)*

## Background Information
Two forces, weathering and erosion, are constantly at work changing the surface of the Earth. Weathering causes rocks to fragment, crack, crumble, or break down. The three general types of weathering are mechanical, chemical, and biological. They usually occur slowly over time and go unnoticed. Mechanical weathering, sometimes called physical weathering, involves the breakdown of rocks and soils through direct contact with atmospheric conditions such as heat, water, ice, and pressure. It does not change the chemical composition of the rocks. Chemical weathering involves the direct effect of atmospheric chemicals in the breakdown of rocks, soils, and minerals. It changes their chemical composition. The third type of weathering is biological weathering caused by plants and animals as they break down the surface of the Earth.

All kinds of rocks weather, but not in the same way or at the same rate. It depends on the composition of the rock, as well as where the rock is located and the conditions to which it is submitted. This activity will look at ways in which mechanical weathering takes place.

Mechanical weathering processes:
- *Heating and cooling*—Changes in temperature can cause the outer layers of rock to peel off. Thermal expansion, as the process is called, usually happens in hot, dry regions where there are maximum daily temperature differences. Exfoliation is the breaking off of rock in curved sheets or slabs.
- *Abrasion*—Rock fragments act as a file and grind the surface of other rocks. Wind or water can carry the rock fragments.
- *Freezing and thawing*—When water freezes in a crack in a rock, it expands and makes the crack bigger. The process is also known as frost wedging. It often widens cracks in sidewalks and causes potholes in streets.

 

**Management**

1. Set up stations. Materials for each station are on the station cards.
2. You may want to set up multiple stations so that students can rotate through them more quickly.
3. Softer rocks such as sandstone, shale, or limestone work best for *Station One.*
4. A large peanut butter jar will work well. Make sure the lid screws on tightly.
5. Two-liter bottles can serve as water sources for the stations. A two-liter bottle with the neck cut off can serve as the water collection container.
6. Use hot tap water for *Station Three.*
7. You will need access to a freezer in which you can place the plastic eggs from *Station Two* overnight.
8. Each student will need a Weathering Ways journal. Cut the pages apart, order them, and staple along the left side.
9. Plastic eggs area available from AIMS (item number 1980).

**Procedure**

1. Explain to the students that they are going to work through stations that deal with some of the different types of weathering.
2. Introduce the class to the weathering stations set up around the room. Distribute the *Weathering Ways* journals.
3. Allow time for students to work through each station.
4. The following day, after they have observed their frozen plastic eggs, discuss what they observed and learned about weathering.

**Connecting Learning**

1. What is weathering?
2. How does moving water play a part in weathering?
3. How does ice play a part in weathering?
4. How do heating and cooling play a part in weathering?
5. What natural forces produce weathering in our state?
6. What evidence do you find that weathering has taken place?
7. What are you wondering now?

 

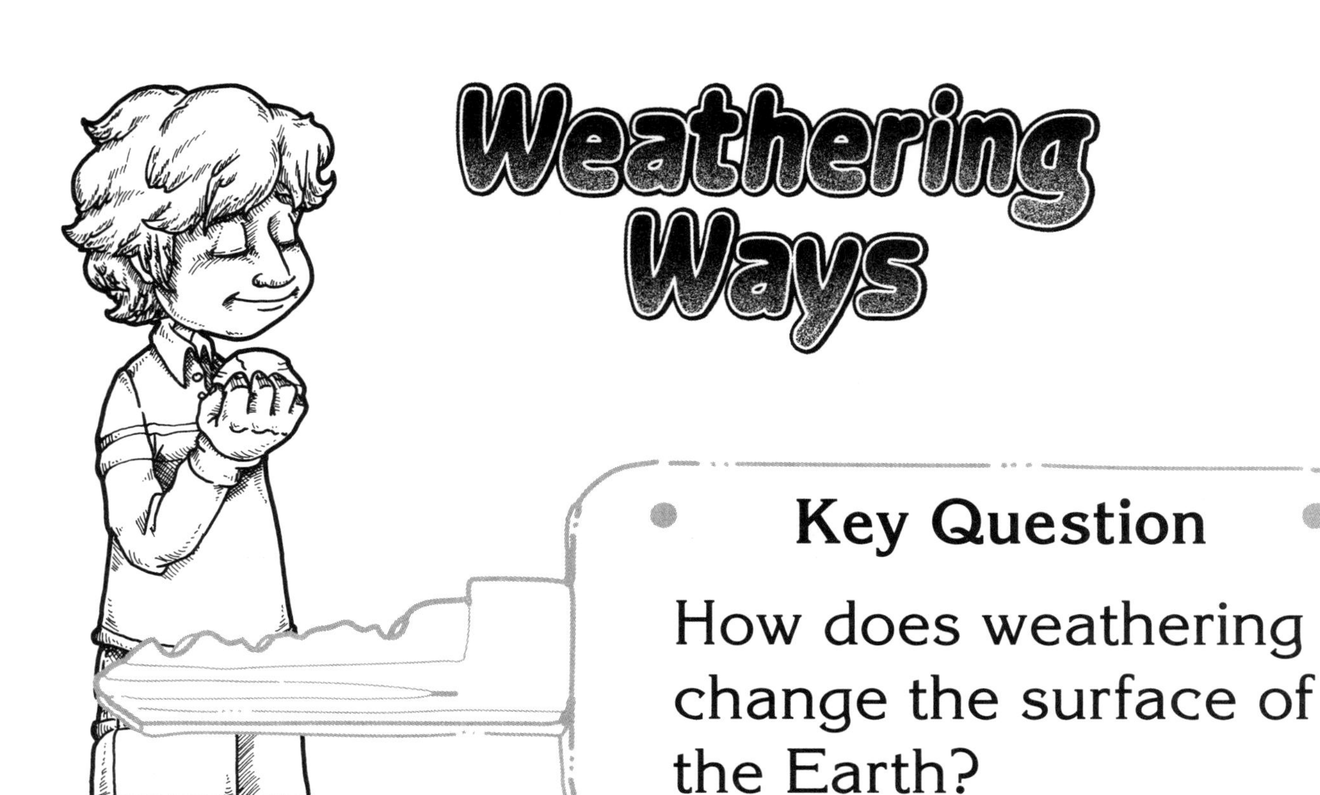

# Weathering Ways

**Key Question**

How does weathering change the surface of the Earth?

## Learning Goals

**Students will:**

- explain the causes of weathering, and
- create models depicting various types of weathering.

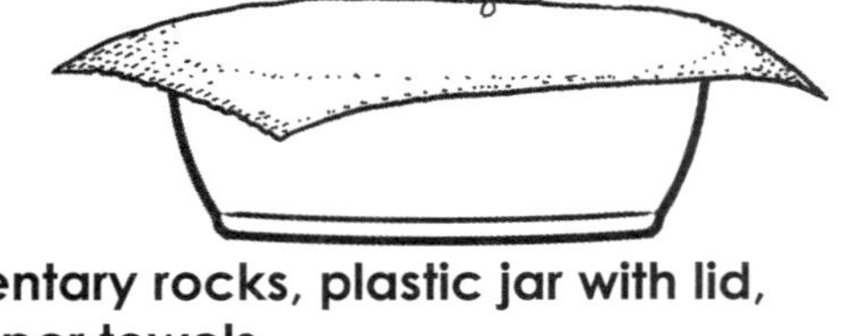

Materials: small broken pieces of sedimentary rocks, plastic jar with lid, water collection container, paper towels

1. Place three small, freshly broken pieces of rock in a plastic jar.
2. Fill the jar about halfway with water.
3. Close the lid of the jar and shake it 1000 times.
4. Remove the rocks and note any changes in their appearance.
5. Cover the water collection container with a paper towel and gently pour the water through it to filter.
6. Answer the questions in your journal.

 

## STATION 2 How does ice weather rocks?

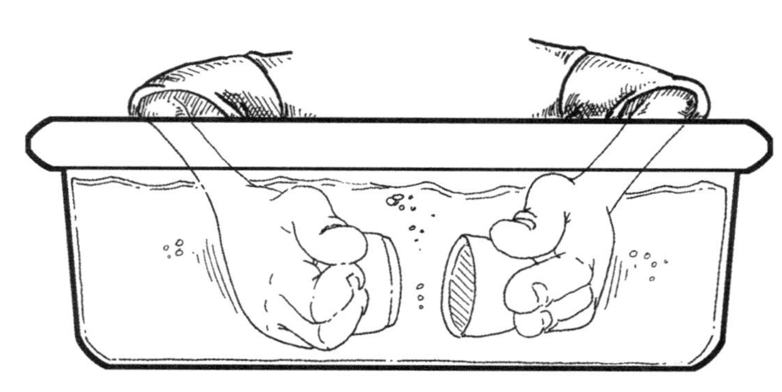

Materials: plastic eggs, water, zipper-type plastic bags, tub of water

1. Place the two plastic egg parts under water and completely fill them with water. Close the egg tightly before lifting it from the water.
2. Place the egg in a plastic bag. Close tightly.
3. Put the bag in a freezer overnight.
4. Observe the next day. Answer the questions in your journal.

## STATION 3 How do heating and cooling weather rocks?

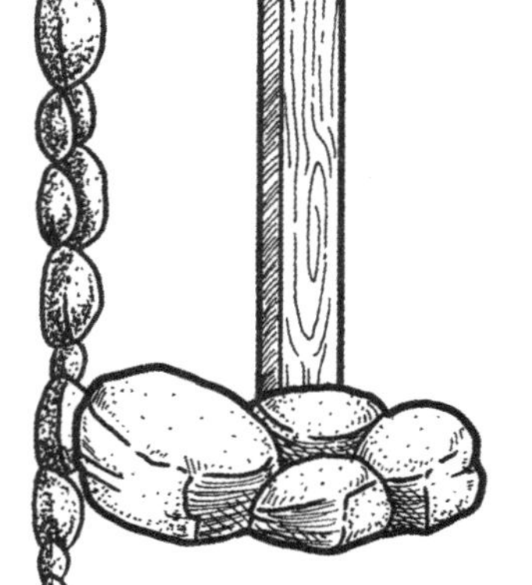

Materials: plastic cup, hot water, ice cubes

1. Fill the cup half full of hot water.
2. Place an ice cube in the water.
3. Answer the questions in your journal.

Weathering
Ways
Station 1
How do streams weather rocks?
What did you do at this station?
Compare the rocks that were shaken with the rocks that were not shaken.
What do you see in the filter paper?
What might happen to rocks in a stream?

 

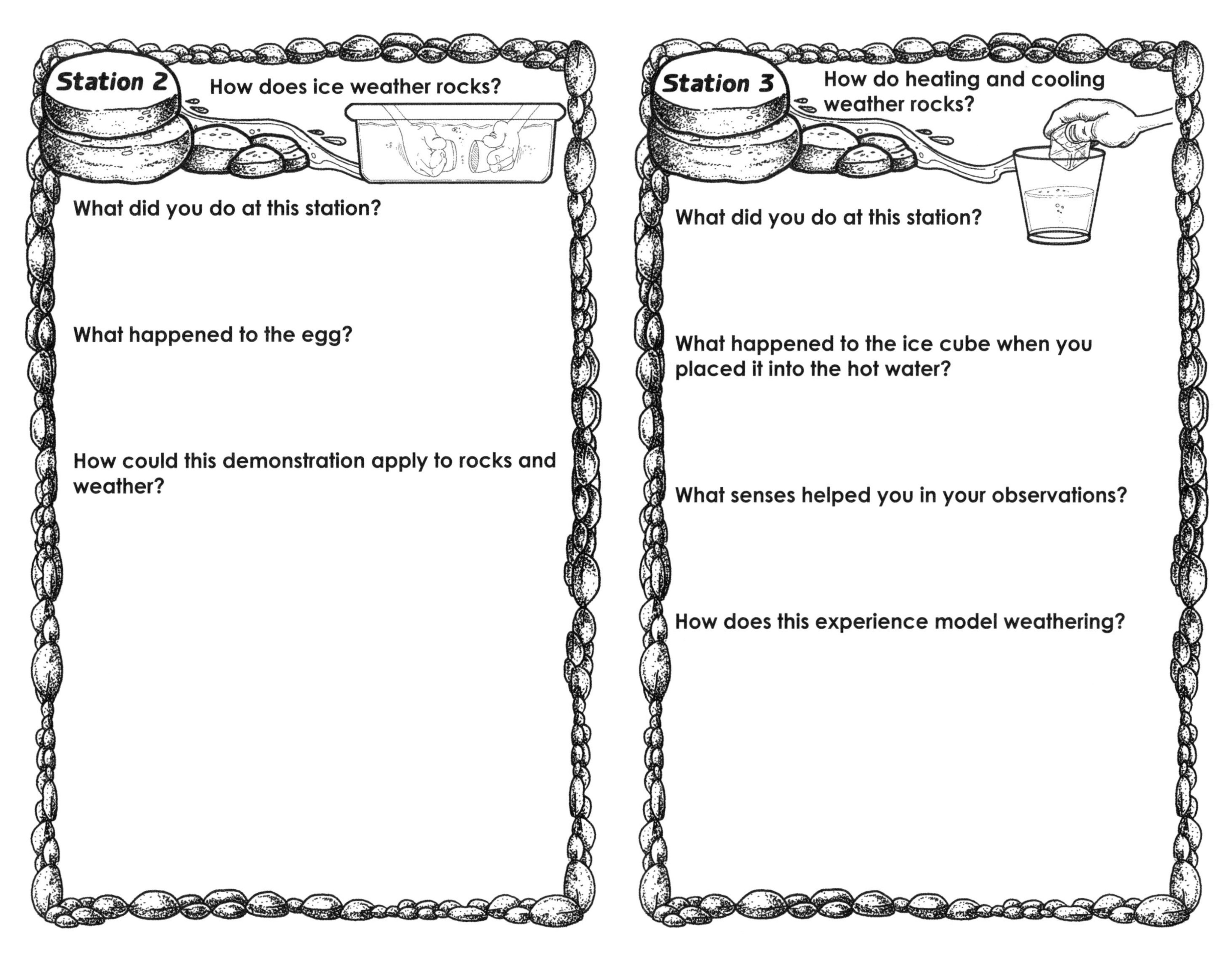

## Station 2

How does ice weather rocks?

What did you do at this station?

What happened to the egg?

How could this demonstration apply to rocks and weather?

## Station 3

How do heating and cooling weather rocks?

What did you do at this station?

What happened to the ice cube when you placed it into the hot water?

What senses helped you in your observations?

How does this experience model weathering?

CONNECTING LEARNING

## Connecting Learning

1. What is weathering?

2. How does moving water play a part in weathering?

3. How does ice play a part in weathering?

4. How do heating and cooling play a part in weathering?

5. What natural forces produce weathering in our state?

6. What evidence do you find that weathering has taken place?

7. What are you wondering now?

**Topic**
Weathering

**Key Question**
What are some ways that plants can change rocks?

**Learning Goals**
Students will:
- model how plants can break down rocks, and
- identify how plants are breaking down rocks in the real world.

**Guiding Documents**
*Project 2061 Benchmarks*
- *People can often learn about things around them by just observing those things carefully, but sometimes they can learn more by doing something to the things and noting what happens.*
- *Change is something that happens to many things.*
- *A model of something is different from the real thing but can be used to learn something about the real thing.*

*NRC Standard*
- *The surface of the earth changes. Some changes are due to slow processes, such as erosion and weathering, and some changes are due to rapid processes, such as landslides, volcanic eruptions, and earthquakes.*

**Science**
Earth science
weathering

**Integrated Processes**
Observing
Comparing and contrasting
Recording
Analyzing

**Materials**
*For each student group:*
plaster of Paris (see *Management 2)*
radish seeds
potting soil
2 plastic cups, 9 oz

*For the class:*
digital camera, optional

**Background Information**
The Earth's surface is subject to change by weathering. Weathering is a general term that is used to describe processes that operate at or near the surface. There are generally three types of weathering: biological, mechanical (physical), and chemical. These processes cause the solid surface of the Earth to dissolve, decompose, and break into smaller pieces. Erosion moves these changed pieces from place to place.

This activity will focus on biological weathering. Biological weathering describes the action of living things on the solid surface of the Earth. Roots of plants prying into the rock can cause them to break apart. Lichen is often found breaking down the surface of the rocks to which it is attached. In this activity, plaster of Paris is used as a model for rocks. The radish seeds will push through and break the plaster of Paris.

**Management**
1. Begin this activity on a Monday.
2. Prepare enough plaster of Paris so that you will be able to pour a thin layer in the plastic cup for each group.
3. *Part Two* needs to be completed after the radish plants have broken through the plaster of Paris.
4. Find some evidence of how plants are changing rocks before you take the students on the field trip. Look for tree roots breaking through sidewalks.

**Procedure**
*Part One*
1. Ask the *Key Question* and state the first *Learning Goal.*
2. Distribute the radish seeds, potting soil, and 9-oz cups to each student group.
3. Direct the students to plant the radish seeds. Have students slightly dampen the soil.
4. Tell the students that you will be pouring a layer of plaster of Paris on the surface of the soil of one of the cups that each group just planted.
5. Have them predict what will happen.
6. Direct the students to place the cups in a sunny, warm place.
7. Have them observe the cups throughout the week and record observations on the first student page.

 

*Part Two*
1. Ask the *Key Question* and state the second *Learning Goal.*
2. Initiate a discussion on what the students observed about the radish plants and the plaster of Paris.
3. Tell the students that they will be taking a field trip around the school ground to find evidence of plants changing rocks. Have them record the evidence they find on the second student page using words and/or pictures. If a digital camera is available, take photos that can be projected for class review and reinforcement.
4. If appropriate, send the page home with students so they can look for evidence of plants changing rocks there.

**Connecting Learning**
1. What are some ways the surface of the Earth can change?
2. Why was it important to make observations?
3. Why did you need to observe two containers?
4. What did the plaster of Paris represent?
5. What real-world examples of weathering were you able to find?
6. What did you learn about weathering? [It breaks rocks into smaller pieces.]
7. What are you wondering now?

 

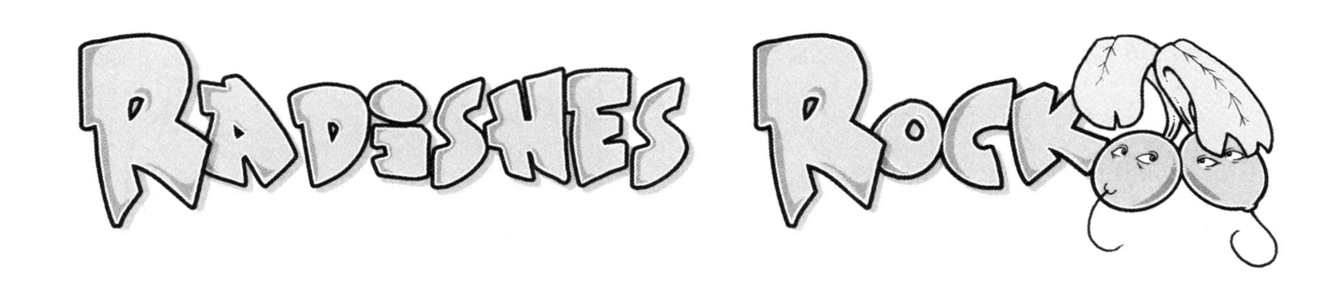


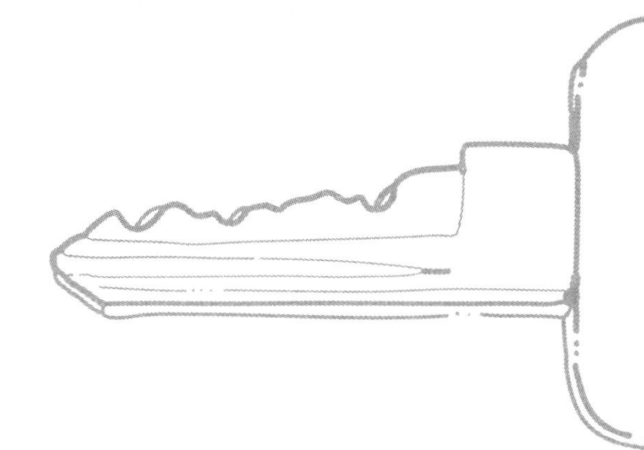

**Key Question**

What are some ways that plants can change rocks?

## Learning Goals

**Students will:**

- model how plants can break down rocks, and
- identify how plants are breaking down rocks in the real world.

 

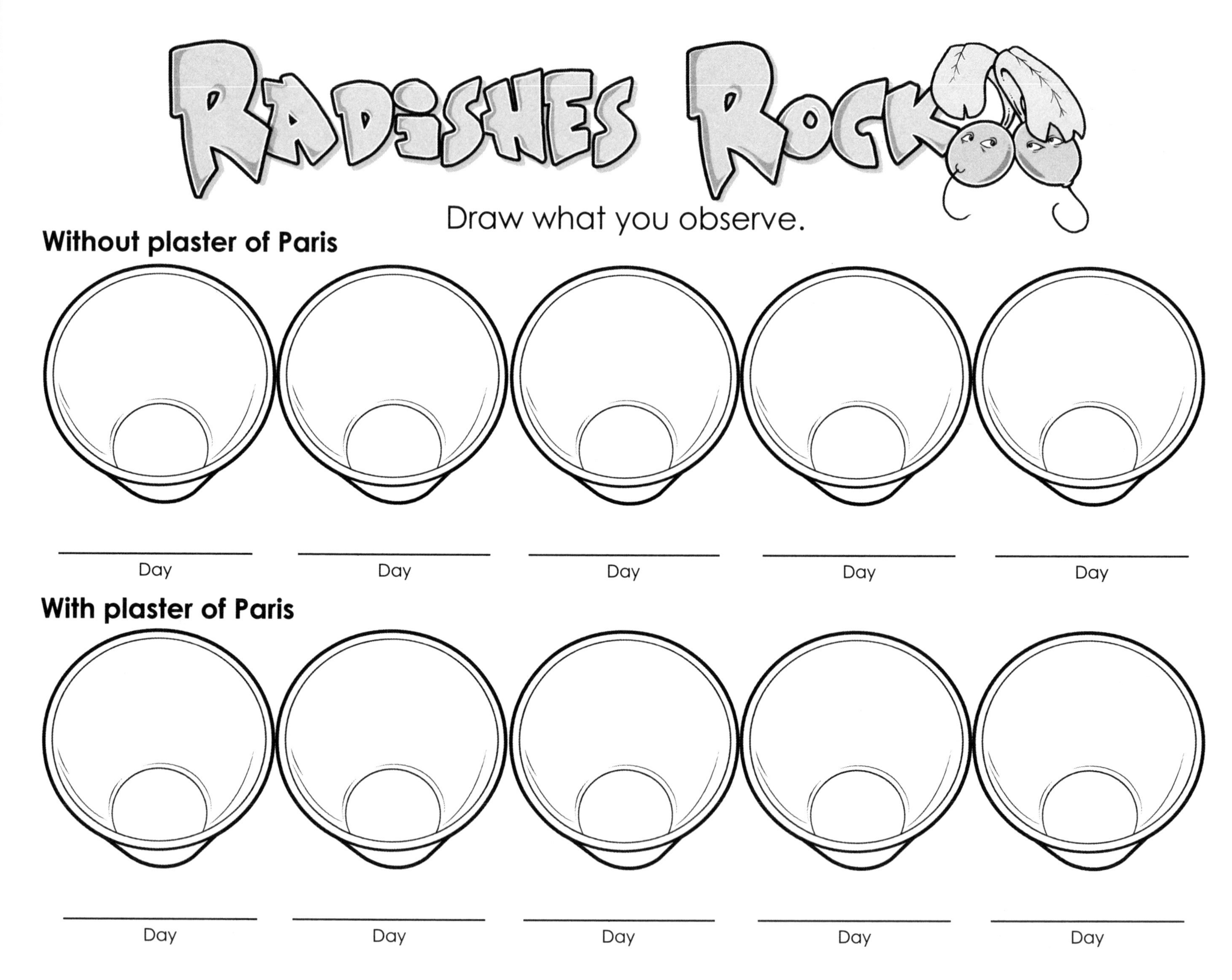
Radishes Rock
Draw what you observe.
Without plaster of Paris
Day
Day
Day
Day
Day
With plaster of Paris
Day
Day
Day
Day
Day

# Radishes Rocks

Evidence of plants changing rocks.

CONNECTING LEARNING

## Connecting Learning

1. What are some ways the surface of the Earth can change?
2. Why was it important to make observations?
3. Why did you need to observe two containers?
4. What did the plaster of Paris represent?
5. What real-world examples of weathering were you able to find?
6. What did you learn about weathering?
7. What are you wondering now?

 

# EROSION

Earth's surface is always changing. Sometimes it changes very quickly. Sometimes it changes so slowly that it is hard to observe.

1

You have learned that weathering is the breaking down of Earth's surface into smaller pieces. *Erosion* is the process that picks up and carries away these pieces.

2

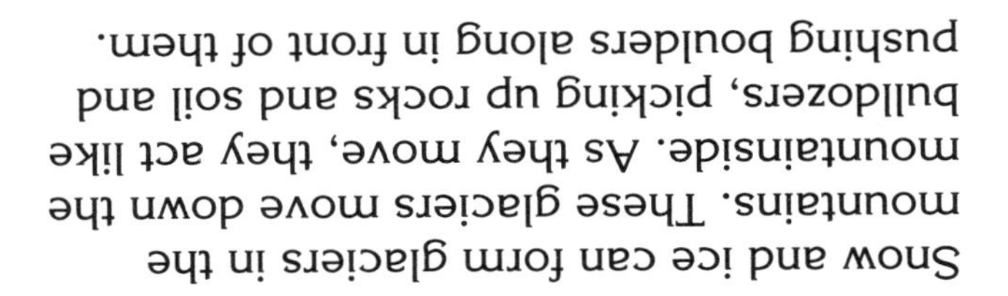
Snow and ice can form glaciers in the mountains. These glaciers move down the mountainside. As they move, they act like bulldozers, picking up rocks and soil and pushing boulders along in front of them.

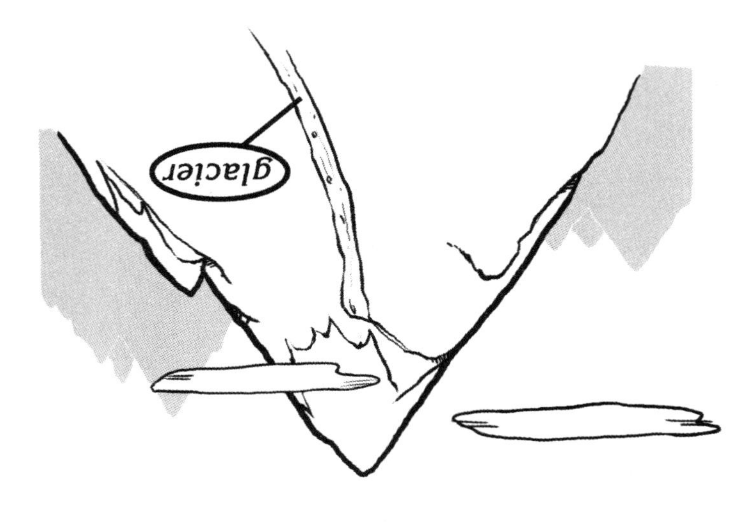


7

Erosion is a natural process, but there are ways we can slow it down. One way is to plant cover crops on bare land or plant trees to slow the speed of the wind. Building terraces, raised rows of dirt, in fields helps to slow the runoff of the rain.

8

How does erosion occur? There are three major tools, or agents, of erosion. These are water, wind, and ice.





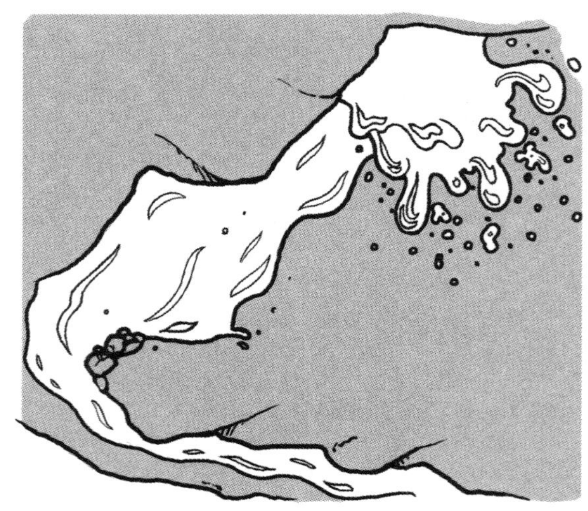

Water is the most powerful agent of erosion. Rain water picks up soil and sand as it runs off. Rivers carry away rocks and soils. Over many years, rivers move enough materials to create valleys and canyons. Along the coastal areas, the waves of the oceans move sand and rocks.



Wind erosion happens mostly along the ground surface. Wind carries off soil and small rocks. You have probably felt the sting of the dirt and sand when it blows against your legs or in your face.

In the 1930s, large amounts of soil were lost when areas of the Plains received little rain. Plants died and could no long anchor the soil when the winds blew. This area and time became known as the Dust Bowl.

DRAWING FROM AN ACTUAL PHOTO, CIRCA 1935

# Changing the Landscape

**Topic**
Canyons/Deltas

**Key Question**
How does water change the surface of the Earth?

**Learning Goals**
Students will:
- model the movement of water in a stream or river,
- observe how the water changes the surface over which it flows, and
- relate this to the formation of canyons and deltas by moving water.

**Guiding Documents**
*Project 2061 Benchmarks*
- *Waves, wind, water, and ice shape and reshape the earth's land surface by eroding rock and soil in some areas and depositing them in other areas, sometimes in seasonal layers.*
- *Change is something that happens to many things.*
- *A model of something is different from the real thing but can be used to learn something about the real thing.*

*NRC Standard*
- *The surface of the earth changes. Some changes are due to slow processes, such as erosion and weathering, and some changes are due to rapid processes, such as landslides, volcanic eruptions, and earthquakes.*

**Math**
Measurement

**Science**
Earth science
  weathering and erosion
  landforms
    canyons
    deltas
  rivers

**Integrated Processes**
Observing
Comparing and contrasting
Collecting and recording data
Relating

**Materials**
*For each stream table:*
  4 disposable lasagna pans
  scissors
  duct tape
  coffee filter
  nail
  sand or dirt (see *Management 2)*

*For each water source:*
  two-liter plastic bottle
  drinking straw
  water
  hole punch
  duct tape

*For each group:*
  pan or container to collect water
  3 heavy-duty yardsticks or wooden laths
  books or blocks
  newspapers
  ruler

*For the class:*
  computer with projection system
  student page, one per student

**Background Information**
The surface of the Earth is constantly changing. Wind, water, and ice are constantly weathering and eroding landforms and creating new ones. Canyons are created in several ways, but one of those is by the erosive force of water as it flows downhill. Deltas are formed at the mouths of rivers as sediment is deposited into larger bodies of water, such as the ocean. This activity uses a stream table to model the erosive force of water and the formation of canyons and deltas.

**Management**

1. A video taking you through the stream table construction process is included on the accompanying CD. You will need to make the stream tables ahead of time. If possible, make one for every group of four to five students.
2. The stream tables can be filled with a variety of Earth materials. You can use sand from the sandbox if you have one on your campus or simply gather dirt from the flowerbeds. Sand can also be purchased at any home improvement store. The pans need not be full, but there should be at least an inch of material in the stream tables.
3. Prepare the water sources for the stream tables ahead of time by following the instructions on the video. Make one water source for each group.
4. Covering the work areas with newspaper will help clean up any drips or spills.
5. When setting up the stream tables, elevate the top end to a height of about 15 cm. Be sure that the end with the drainage holes is in the container for collecting water.

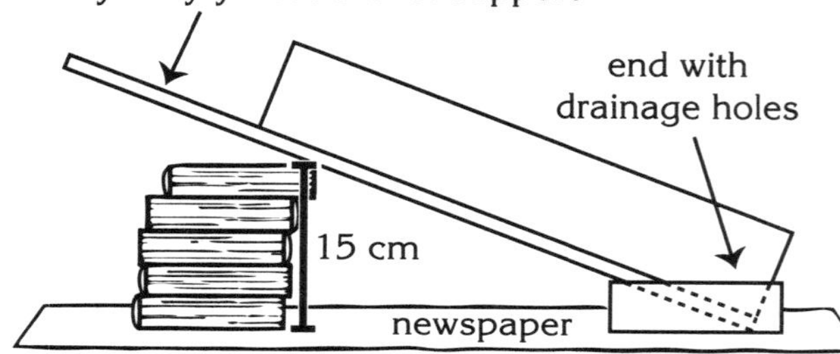


6. To facilitate the setup of multiple stream tables, have the empty stream tables along with the other necessary materials in a location where students can come and collect them. Put the sand/dirt in large plastic bags so that students can pour it into their stream tables once they are set up rather than trying to carry the stream tables full of sand.
7. You will need a computer with a projection system in order to show students the video of a stream table in action as well as the pictures of canyons and deltas.

**Procedure**

1. Ask the *Key Question*, "How does water change the surface of the Earth?" and have students share what they already know.
2. Discuss how some processes take a long time to happen and happen on such a large scale that you can't observe them directly—you have to use models. Tell students that they will be using stream tables to model how the action of water forms canyons and deltas.
3. Divide students into groups and have each group cover its work area with newspaper.
4. Instruct groups to gather the necessary materials, and show them how to place their stream tables on the books so that the ends are raised to a height of 15 cm. Be sure that they have the yardsticks supporting the bottom of the stream table and the pan at the end to collect water.
5. When groups have their stream tables set up correctly, instruct them to add the sand/dirt to their stream tables. Inform them that they need to smooth the surface of the sand and make sure that it's about the same depth all the way through. Tell them to fill only the top three-quarters of the pan, leaving about 20 cm at the bottom of the pan with no sand.

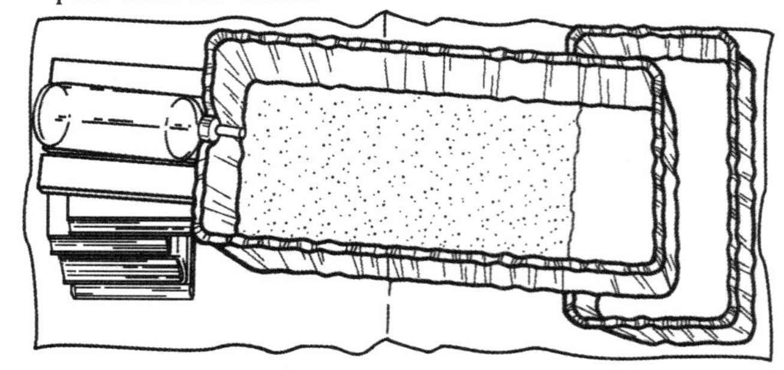

6. Distribute the student page and have students complete the *Before* pictures, including the measurements of where the sand ends relative to the bottom of the pan.
7. Show students how to bend a small indentation in the top of the stream table for the neck of the two-liter bottle. Demonstrate how the water does not come out of the straw when the other hole in the duct tape is covered. Have them set their water sources in place and then remove their fingers to start the water flowing.
8. As students observe their stream tables in action, encourage them to pay attention to how the water behaves, pointing out places where it creates features you might find in a real river, such as oxbows or islands. When the water in the bottles ceases to flow on its own, students can tip the ends of the bottles up to get out most of the remaining water.
9. Encourage students to complete the student page by recording their observations and identifying the landforms created. Be sure that they measure the distance the sand/dirt is from the bottom of the pan and identify the size of any new deposits that may have formed where there was no sand/dirt previously.
10. Have students move around the room and observe other groups' stream tables. Discuss similarities and differences among the groups. If there are any stream tables with especially nice examples of deltas or canyons, point these out to the class.

 

11. Show the provided video of a stream table in action to students and make comparisons between what happens in the video and what they observed in their own stream tables.
12. Project the pictures of canyons and deltas for the class to see. Discuss how these real-world examples compare to the deltas and canyons that were created in their stream tables.

**Connecting Learning**

1. What are some ways that water changes the surface of the Earth?
2. How did your stream table model what happens with water in the real world?
3. What landforms did you observe in your stream table?
4. How did the appearance of your group's stream table compare to that of others in the class? How can you explain the differences?
5. How did your model canyons and deltas compare to real canyons and deltas?
6. What are you wondering now?

**Extensions**

1. Experiment with the slope of the hill and how that affects the changes made by the water.
2. Explore different Earth materials in the stream table to compare how they behave.
3. Smooth out the sand/dirt and repeat the experiment while it is still wet. How do the results compare?

# Changing the Landscape

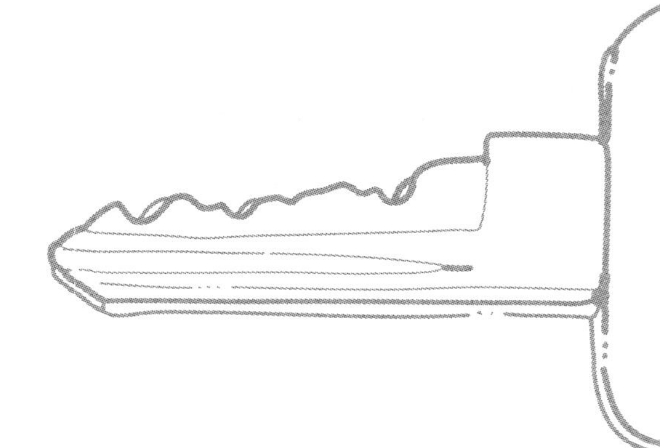

**Key Question**

How does water change the surface of the Earth?

## Learning Goals

- model the movement of water in a stream or river,
- observe how the water changes the surface over which it flows, and
- relate this to the formation of canyons and deltas by moving water.

 

# Changing the Landscape

Draw your stream table before you start the water flowing. Include measurements that tell where the sand stops at the bottom of the pan.

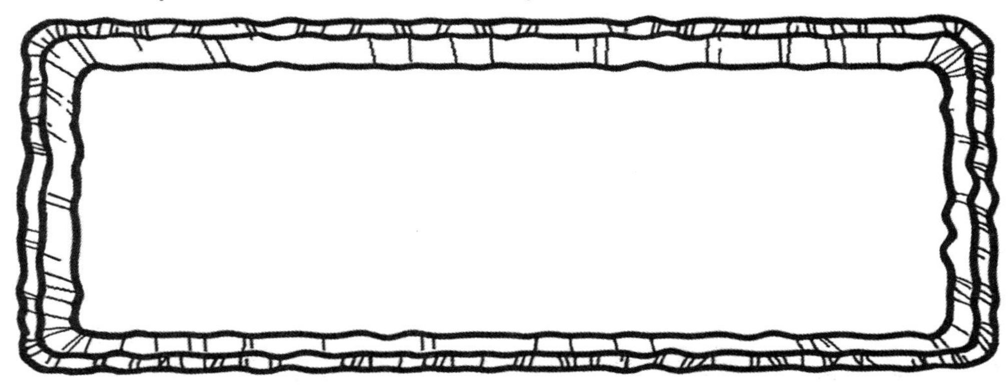

Start the water flowing. Make careful observations of the path of the stream and the movement of the sand. Draw how your stream table looked after the water stopped flowing. Be sure to measure the movement of the sand and draw in any paths the water took.

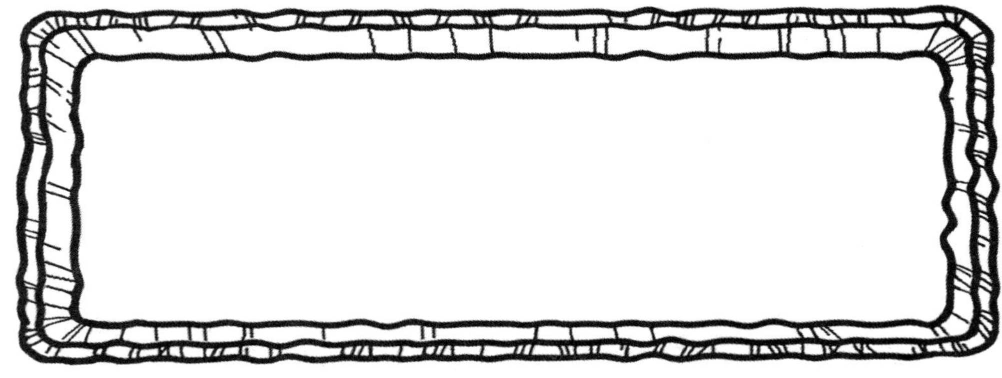

Describe what you observed as the water flowed in your stream table.

How did the water change the surface of the "Earth" in your stream table?

What landforms were created?

 

# Changing the Landscape

CONNECTING LEARNING

## Connecting Learning

1. What are some ways that water changes the surface of the Earth?
2. How did your stream table model what happens with water in the real world?
3. What landforms did you observe in your stream table?
4. How did the appearance of your group's stream table compare to that of others in the class? How can you explain the differences?
5. How did your model canyons and deltas compare to real canyons and deltas?
6. What are you wondering now?

**Topic**
Water erosion and deposition

**Key Question**
How does a river's slope affect the land through which it flows?

**Learning Goals**
Students will:
- learn that a river's slope influences the amount of erosion that takes place, and
- realize that dams change the distribution of sediment along a river.

**Guiding Documents**
*Project 2061 Benchmark*
- *Waves, wind, water, and ice shape and reshape the earth's land surface by eroding rock and soil in some areas and depositing them in other areas, sometimes in seasonal layers.*

*NRC Standard*
- *The surface of the earth changes. Some changes are due to slow processes, such as erosion and weathering, and some changes are due to rapid processes, such as landslides, volcanic eruptions, and earthquakes.*

**Math**
Line graph

**Science**
Earth science
  erosion
  deposition

**Social Science**
Geography
  U.S. rivers

**Integrated Processes**
Observing
Recording data
Comparing and contrasting
Relating
Drawing conclusions

**Materials**
Colored pencils
River pictures, included
Student pages

**Background Information**
Rivers typically begin in hills or mountains as small gullies of melted ice and snow. The water flows downhill by gravity, eroding soil and bits of rock along its path. This sediment is carried downstream and is eventually deposited along lower banks and at the mouth of the river where it may form a delta.

Whether rock is eroded or sediment is deposited varies with the velocity or speed of the water. The faster the river flows, the more kinetic energy it has to scrape and cut away rocks. The slower the river flows, the lower its energy and the more the sediment it is carrying settles on its banks.

The velocity of water varies with gradient, or slope, and the size of the channel. Water flowing down a steep slope has higher velocity than water flowing down a gentler slope. Water in a narrow channel flows more quickly than water in a wide channel.

Because the Mississippi River begins at a relatively low elevation and travels over 2000 miles to reach sea level, its slope is quite gradual. This slope, coupled with a wide channel, results in relatively slow-moving water—typically less than one mile per hour on the lower Mississippi. The river does not have the kinetic energy to carve out canyons. The Mississippi's large volume, however, allows it to carry and deposit sizeable amounts of sediment; its delta is over 100 miles long!

In contrast, the Colorado travels a shorter distance from an elevation over 10,000 feet, so its slope is relatively steep. The steep slope causes the water to flow more quickly. Over geological time, the water cut into the rocks of the Colorado Plateau and, along with wind and weathering, helped form the mile-deep Grand Canyon.

**Management**
1. Each student will need two pencils of different colors.
2. Photographs of the Mississippi and Colorado Rivers are included on the accompanying CD. It is best if these are displayed using a computer with a projector.
3. The graph of the two rivers, called a long profile, shows how the slope of each river changes along its length. Table distances and elevations have been rounded.
4. Encourage students to work in pairs when plotting points and discussing possible conclusions.

**Procedure**

1. Set the following scene for students: You've just finished dinner. Your plate still has bits of food on it. You hold the plate at an angle under the running water of the kitchen faucet. What happens? [The water washes most of the food bits away.]
2. Explain that, over a long time, the flowing water in rivers can do something similar to land. Tell students that they will look at two large rivers to learn how a river's slope affects the land through which it flows.
3. Show the pictures of the Mississippi River. Ask students what they observe about the river and the land around it. [It is typically wide, muddy, and the surrounding land is relatively flat.]
4. Distribute the Mississippi River page. Have students trace the path of the river from its source to the Gulf of Mexico, referring to the table to check elevation changes along the way.
5. Give students the graph page. With a colored pencil, have them draw a profile of the Mississippi using the data in the table. Invite students to analyze the scale on the graph. [The horizontal dots represent 25-mile increments and the vertical dots represent 100-foot increments.]

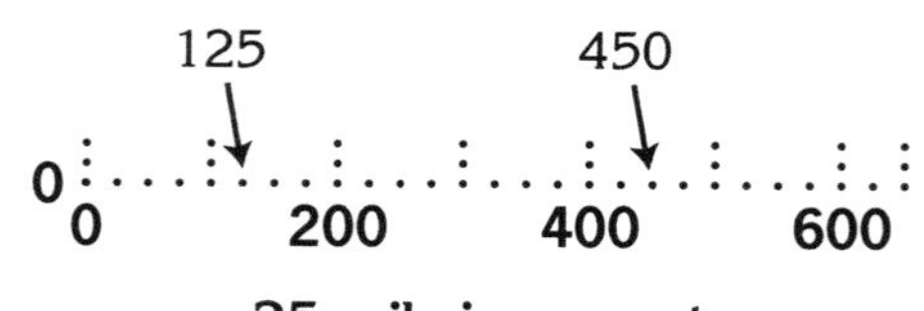


6. Show the pictures of the Colorado River and distribute the Colorado River page. Ask students what they observe about the river and the land around it. [The river is narrower than the Mississippi. It is at the bottom of deep canyons, not in the middle of flat land.]
7. Have students trace the path of the Colorado River, notice elevation changes between specific points, and then plot its profile with a different colored pencil. Direct them to label each profile with the river's name.
8. Guide students in comparing the two rivers using the pictures and the graph.
9. Distribute the *Flooding the Colorado* page and have students read how dams can affect sediment distribution.
10. Show students the pictures of the Glen Canyon Dam and the Mississippi River Delta. Ask them to look for evidence of deposition in the two pictures. Be sure they recognize what they are looking at in the picture of the delta (the gulf of Mexico as seen from space). Tell students that the Glen Canyon Dam picture was taken just after the 2008 high-flow release.

**Connecting Learning**

1. In what ways can a river change the land around it? [erode the banks, carve canyons, carry and build up sediment deposits on the lower river, form deltas, etc.]
2. What are some observations you can make from the river profiles and pictures? [Rivers flow downhill and empty into the ocean at sea level. Rivers lose most of their height on the upper part of the river; slopes usually decrease downstream. The steeper the slope of the river bed, the more land it seems to erode. Rivers carry soil and eroded rock downstream. Deposition can happen along the path of a river and at the mouth of a river.]
3. Name a past event on the Colorado River that affects the present. [forming of the Grand Canyon, building dams, flood release] Describe one of the effects this event has had.
4. What are most of the river systems in your state like? Where can we see evidence of erosion and deposition?
5. What are you wondering now?

**Extensions**

1. On a large map of the United States, have students look for rivers that feed into the Mississippi or Colorado Rivers.
2. Choose another river such as the Rio Grande or Snake River. Suggest students gather elevation data for specific points along the river and graph its profile. For example, the Rio Grande begins in southwestern Colorado at about 13,000 feet.
3. Research other examples of river erosion such as Mammoth Cave and Niagara Falls.
4. Challenge students to search for other rivers that either support or refute one of their conclusions. For instance, find a river that does not empty into an ocean.
5. Encourage students to design an investigation to show that water flows faster on a steep slope than on a gentle slope.

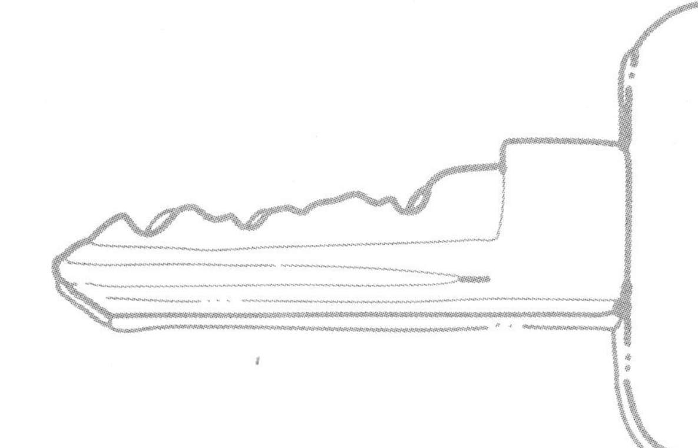

## Key Question

How does a river's slope affect the land through which it flows?

## Learning Goals

**Students will:**

- learn that a river's slope influences the amount of erosion that takes place, and
- realize that dams change the distribution of sediment along a river.

Colorado River 1

Colorado River 2

Colorado River 3

Colorado River 4

Colorado River 5

Colorado River 6

Glen Canyon Dam

Mississippi River 1

Mississippi River 2

Mississippi River 3

Mississippi River 4

Mississippi River 5

Mississippi River 6

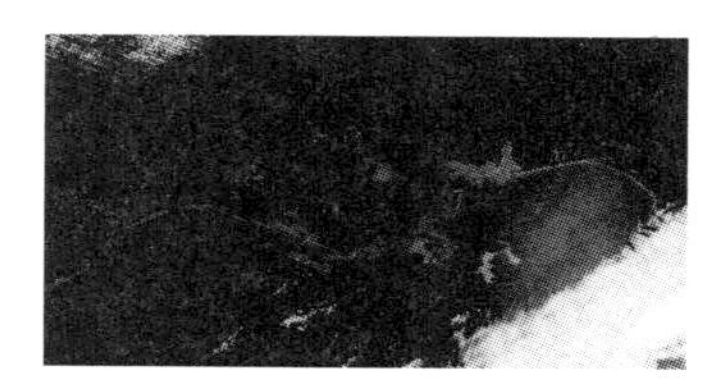
Mississippi River Delta

 

# River Run

## Mississippi River

| Location | Distance (miles) | Elevation (feet) |
|---|---|---|
| Gulf of Mexico | 0 | 0 |
| New Orleans, LA | 125 | 20 |
| Vicksburg, MS | 450 | 175 |
| Memphis, TN | 750 | 260 |
| Cairo, IL | 1000 | 300 |
| St. Louis, MO | 1175 | 470 |
| Dubuque, IA | 1575 | 620 |
| Minneapolis, MN | 1850 | 830 |
| Grand Rapids, MN | 2075 | 1300 |
| Lake Itasca, MN | 2340 | 1475 |

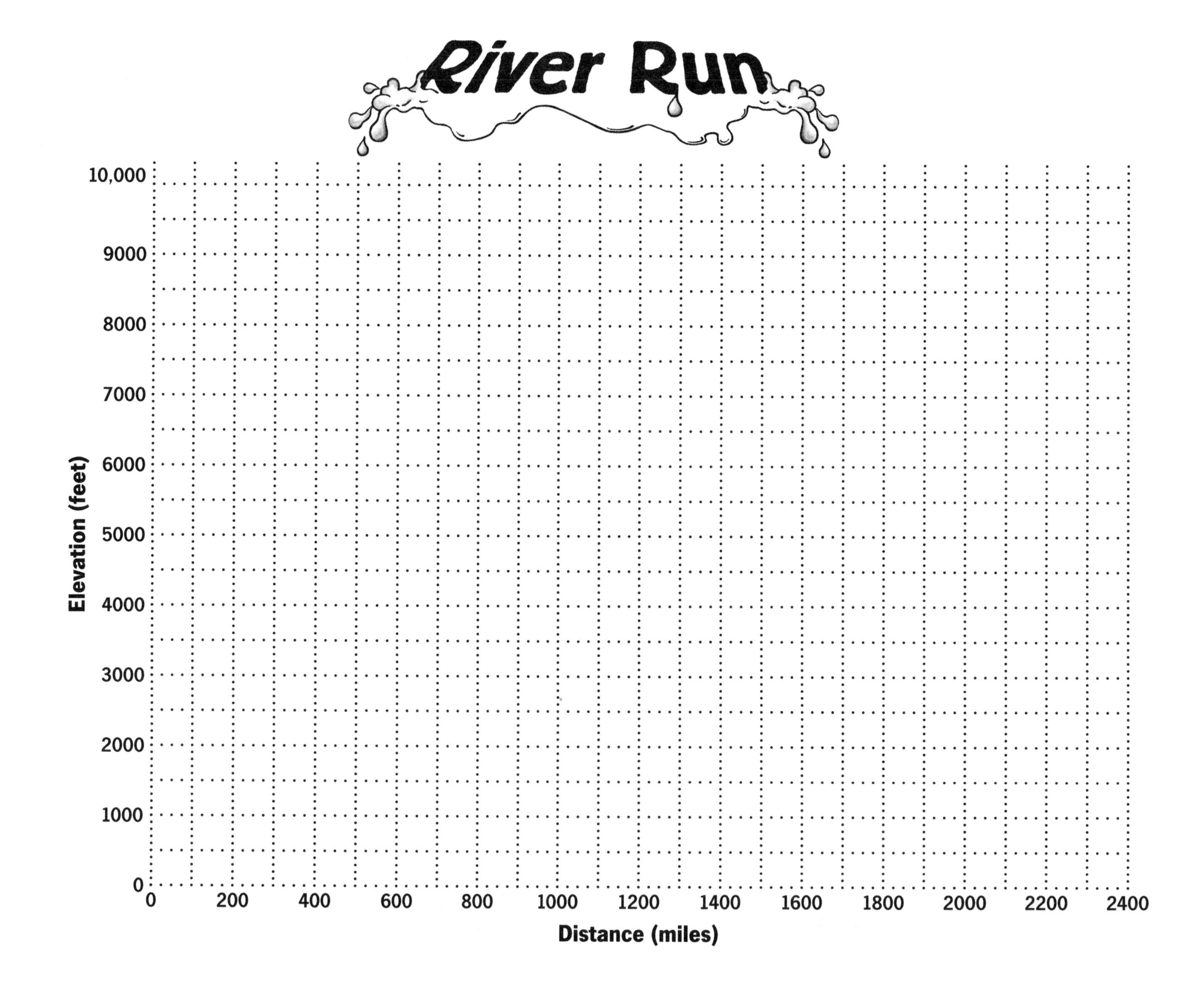
River Run
Elevation (feet)
10,000
9000
8000
7000
6000
5000
4000
3000
2000
1000
0
0
200
400
600
800
1000
1200
1400
1600
1800
2000
2200
2400
Distance (miles)

 

# River Run

## Colorado River

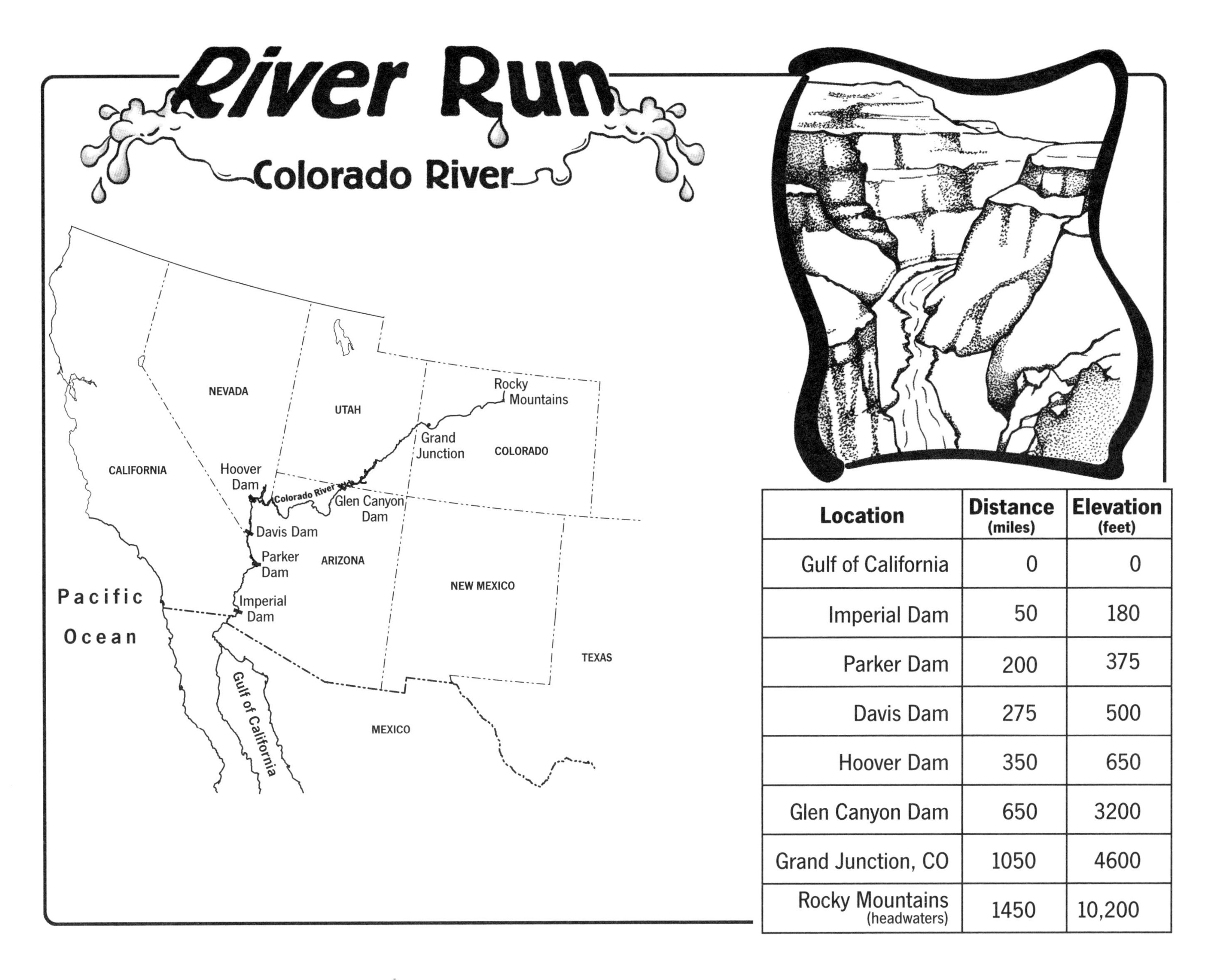


| Location | Distance (miles) | Elevation (feet) |
|---|---|---|
| Gulf of California | 0 | 0 |
| Imperial Dam | 50 | 180 |
| Parker Dam | 200 | 375 |
| Davis Dam | 275 | 500 |
| Hoover Dam | 350 | 650 |
| Glen Canyon Dam | 650 | 3200 |
| Grand Junction, CO | 1050 | 4600 |
| Rocky Mountains (headwaters) | 1450 | 10,200 |

# Flooding the Colorado

Small bits of soil and rock used to give the Colorado River a muddy color. This sediment settled and formed a rich delta in the lower river valley. Every once in awhile, flash floods would scrape parts of the Grand Canyon clean and deposit fresh sand along the river's beaches.

The Hoover Dam was built in 1936 to help control flooding and erosion, provide irrigation water, and generate electrical power. Now there are more than 20 dams along the Colorado River, including the Glen Canyon Dam, completed in 1964.

How have the dams changed the river? Since much of the water is now sent to irrigation canals, the river is often dry before it reaches the Gulf of California. The temperature of the water has become cooler. The sediment collects behind the dams, leaving the river below them clear. Without sediment to keep them filled, beaches at the edges of the river erode. Some of the fish and plants that lived here find it hard to survive in the changed habitat.

To restore beaches and save animal and plant life, much of this sediment would need to be distributed below the dam. This has happened three times—in 1996, 2004, and 2008. Each time, a large amount of water was released from the Glen Canyon Dam, rebuilding beaches with hundreds of thousands of tons of sediment. These *high-flow releases,* as they are called, may become a regular part of the plan for managing the river.

For more information about high-flow releases, before-and-after photographs, and video, visit the USGS's Grand Canyon Monitoring and Research Center website (http://www.gcmrc.gov/). USGS videos of the 2008 high-flow release can also be found on YouTube (http://www.youtube.com/profile?user=ocweb).

## Connecting Learning

1. In what ways can a river change the land around it?
2. What are some observations you can make from the river profiles and pictures?
3. Name a past event on the Colorado River that affects the present. Describe one of the effects this event has had.
4. What are most of the river systems in your state like?
5. What are you wondering now?

**Topic**
Canyons

**Key Question**
How are canyons formed?

**Learning Goals**
Students will:
- form landmasses from clay,
- carve the clay with a knife to model the weathering and erosion of land by rivers, and
- observe the canyon-like structure of the resulting clay formation.

**Guiding Documents**
*Project 2061 Benchmarks*
- *Waves, wind, water, and ice shape and reshape the earth's land surface by eroding rock and soil in some areas and depositing them in other areas, sometimes in seasonal layers.*
- *Change is something that happens to many things.*
- *A model of something is different from the real thing but can be used to learn something about the real thing.*

*NRC Standard*
- *The surface of the earth changes. Some changes are due to slow processes, such as erosion and weathering, and some changes are due to rapid processes, such as landslides, volcanic eruptions, and earthquakes.*

**Science**
Earth science
weathering, erosion
canyons

**Integrated Processes**
Observing
Comparing and contrasting
Modeling
Applying

**Materials**
Clay (see *Management 2)*
Cylinders (see *Management 3)*
Waxed paper (see *Management 4)*
Plastic knives
Colored pencils
Projection device
Internet access
Student page

**Background Information**
Palo Duro Canyon is located in the panhandle of Texas, just south of Amarillo. The canyon is 120 miles long, up to 20 miles wide, and its deepest area is about 800 feet. Palo Duro Canyon is the second largest canyon in the US; the Grand Canyon is the largest. *Palo Duro* is Spanish for *hard wood*. There is an abundance of juniper and mesquite trees in the area.

Geologists predict that the oldest layers of rock date back 250 million years. Most of the rocks in the canyon are sedimentary—gypsum, sandstone, siltstone, and shale.

Wind, water, and ice are agents of weathering for this canyon. It was formed mainly by water erosion by the Prairie Dog Town Fork of the Red River. This area of Texas gets snow during the winter, so ice wedging also weathers the rock. Wind pits the surface of the soft sedimentary rock, and water washes the sediments away.

Two very evident features of weathering and erosion are the hoodoos and the Spanish Skirts. Hoodoos are towering formations with a capstone on top. The Lighthouse is the most noted hoodoo in the canyon. The capstones of the hoodoos are of harder rock. The softer rock underneath more easily wears away. Spanish Skirts are the triangular-looking formations that show multiple layers of colorful rocks. (The Lighthouse and Spanish Skirts are pictured on the homepage of the website you will be using.)

In this activity, students will model the weathering and erosion of rock by using the wide part of the blade of a plastic knife to carve away at layers of clay. They will learn that the top layer of rock is the youngest, the bottom layer is the oldest. This is the Law of Superposition and is relevant to undisturbed layers of rock. As students carve through their models, they will keep track of the sediments that are removed with each carving, noting that they move through the different layers as they cut deeper and deeper. By comparing their models to the photos they will observe, they should be able to relate the colors of their clay layers to the multicolored rock layers seen in the photos.

**Management**
1. Students should work in groups of three or four.
2. Each group will need three or four colors of clay. A generous one-inch cube of clay is sufficient for each color.
3. Students will need to roll out their cubes of clay into flat ovals. Cans filled with fruits, vegetables, or soups work well. Any cylinder that can be rolled using moderate pressure will work.

 

4. Each group will need a sheet of waxed paper that is between 12-18 inches in length.
5. Set clay cubes, waxed paper, plastic knives, and cans in an area that is easily reached by students.
6. Be prepared to show students how to roll out and layer the clay.
7. You will need Internet access and a projection device to show pictures of Palo Duro Canyon. Preview the photos found on the site listed in *Internet Connections* to choose the pictures you wish to show students.

**Procedure**

*Part One*

1. Ask students if they have ever visited a canyon. Invite those who have to share their experiences.
2. Ask the *Key Question* and state the *Learning Goals.*
3. Explain that Earth's surface is covered by layers of rock. They are going to use different colors of clay to represent the different layers of rock.
4. Show the students how to roll out a cube of clay by placing it on one end of a sheet of waxed paper and then covering the top of the clay with the extra length. Use a can to roll the clay to make a shape that is about 4 inches by 3 inches. The clay shape will be thin. Project the page of directions so students can follow them.
5. Invite students to gather their materials. Direct them to roll out all the colors of clay you have provided.
6. Once all the layers have been rolled out, have the students stack them and use the can to apply pressure to bind them together. Students can make a bend in their layered clay to add some elevation contour, if desired.
7. One at a time, have the students in the group use the broad side of the knife blade to carve a river going through the rock layers. They should only scrape the surface of the clay and not try to cut through it. At each carving, have students remove the sediments (the clay) found on the knife and place them in an orderly fashion on the waxed paper. The sequence of these sediments will show them the progressions of the colors of clay that indicate different rock layers.
8. When the carvings reach the final layer, have students stop and make observations of the canyon layers.
9. Distribute the student page and allow time for students to complete *Part One.*

*Part Two*

1. Inform the students that you are going to show them pictures of Palo Duro Canyon that is located in the panhandle of Texas near Amarillo and Lubbock.
2. Show and discuss the photos you have chosen, describing the processes of weathering and erosion and the different formations found in the canyon. See *Background Information.*
3. Have students complete *Part Two* of the student page.
4. Conclude with a discussion on how canyons are formed.

**Connecting Learning**

1. How is our model like what happens in nature? [Rocks are in layers. Rivers cut through rocks like our knives carved through clay. Each cut makes the canyon deeper. Sediments are moved through the canyon. Etc.]
2. How is our model different? [Rocks are harder than our clay. It takes many, many years for a canyon to be carved. Etc.]
3. Describe the sequence of the sediments of clay that you removed. [They started with the top color of clay, moved to the second layer, and then on down. Some sediment clumps had more than one color in them.]
4. Which rock layer was the oldest? How do you know? [The rock layer on the bottom was the oldest. We put that layer down first.]
5. What caused the Palo Duro Canyon to form?
6. What indications were there that sediments were being eroded? [The water was red because of the sediments it picked up as it flowed over the land.]
7. What other canyons have you heard of? How were they formed?
8. What are you wondering now?

**Internet Connections**

*Palo Duro Canyon*

http://www.palodurocanyon.com/

Select *Photos* found in the top menu bar. Select photos that show various canyon formations and erosion. Be sure to show pictures of hoodoos, describing the weathering process (see *Background Information).* Point out the red sediments being eroded away by the water. The winter pictures can be used to discuss ice wedging as a weathering agent.

## Key Question

How are canyons formed?

## Learning Goals

**Students will:**

- form landmasses from clay,
- carve the clay with a knife to model the weathering and erosion of land by rivers, and
- observe the canyon-like structure of the resulting clay formation.

# CARVING CANYONS

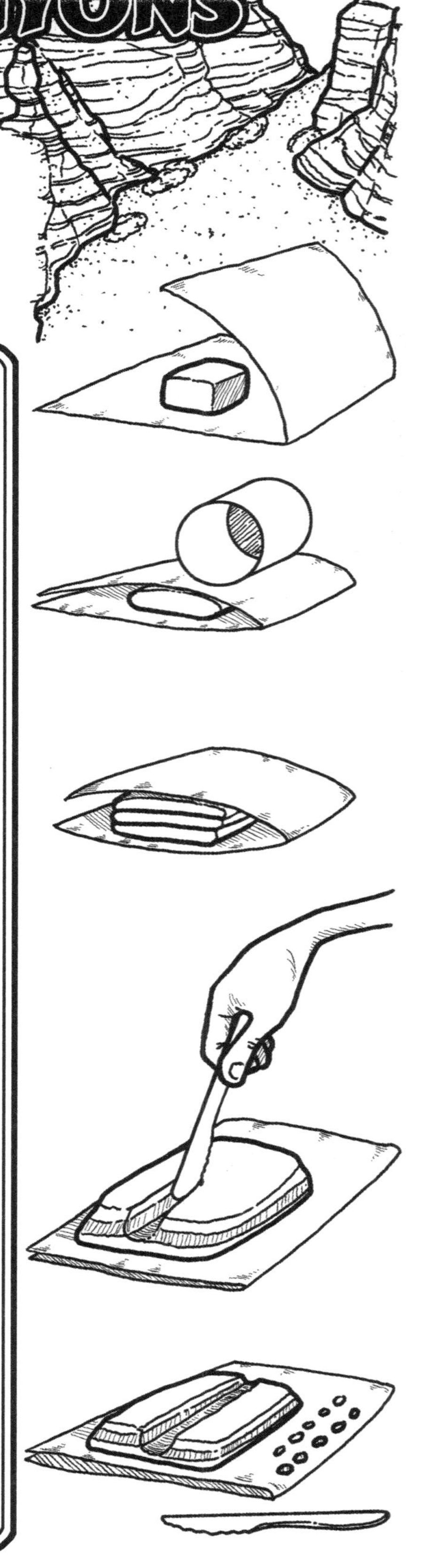

**You will need:**

3 or 4 clay cubes, different colors
1 sheet of waxed paper
1 plastic knife
1 can

**Do this:**

1. Place a cube of clay on one end of the waxed paper. Fold the waxed paper over the top of the clay.
2. Use a can to roll the clay into a thin layer. It should make a blob about 4 inches by 3 inches.
3. Do the same for all your cubes of clay.
4. Stack the rolled out layers of clay. Fold the waxed paper over the top and roll the stack to make them stick together.
5. Uncover the stack of clay layers.
6. Let each group member carve a "river" going through the clay. (Just scrape the surface. Don't try to cut all the way through.)
7. With each carving, remove the clay from the knife and put it on the waxed paper. Keep these "sediments" in order as they are removed so you can see the sequence of the colors.
8. Continue carving until you reach the bottom layer of clay.

 

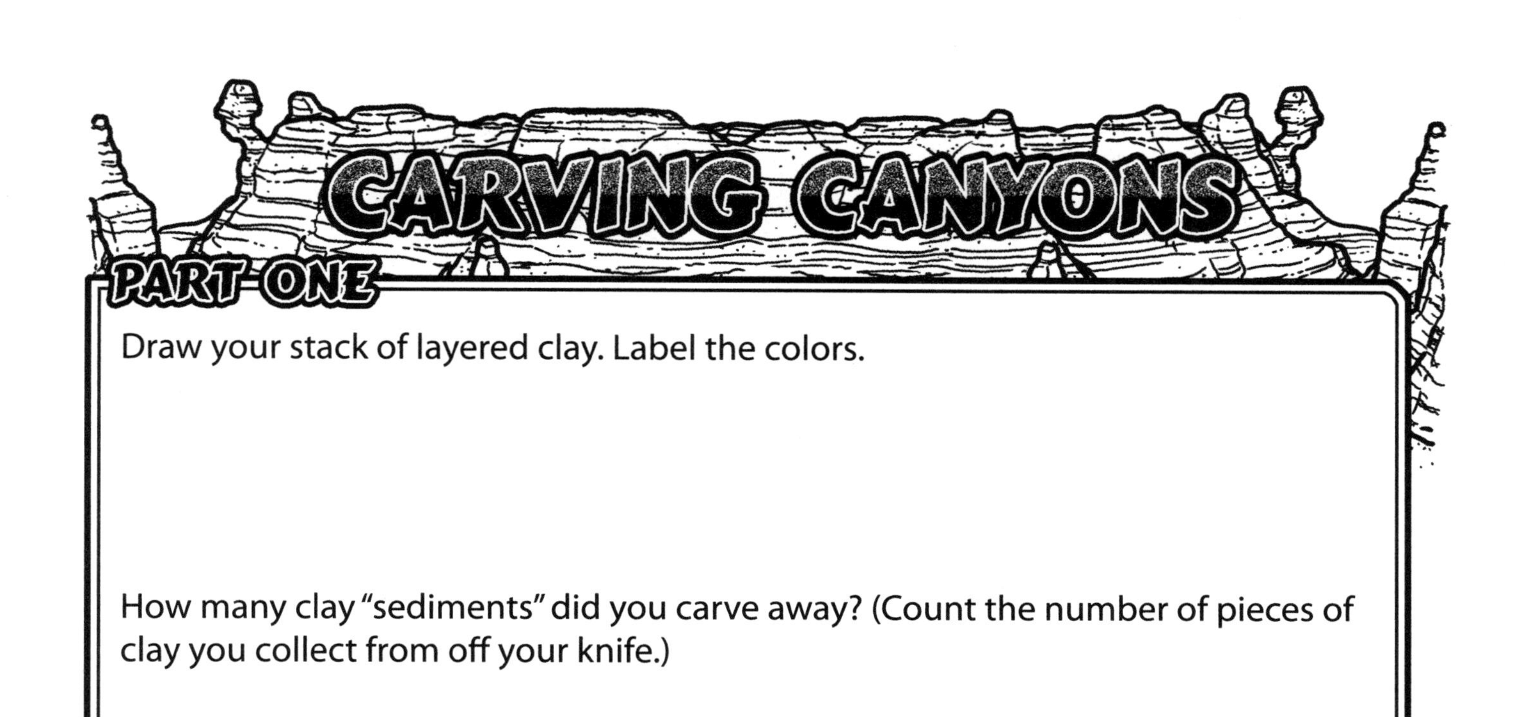


Draw your stack of layered clay. Label the colors.

How many clay "sediments" did you carve away? (Count the number of pieces of clay you collect from off your knife.)

If each of these required 10 million years of river erosion, how many years did it take to form your canyon?

Draw your stack of clay after you have carved your canyon.

List three observations of your carved canyon.

## PART TWO

How is your model canyon like the pictures of the Palo Duro Canyon?

How is it different?

What caused the weathering and erosion that formed the Palo Duro Canyon?

## Connecting Learning

1. How is our model like what happens in nature?

2. How is our model different?

3. Describe the sequence of the sediments of clay that you removed.

4. Which rock layer was the oldest? How do you know?

5. What caused the Palo Duro Canyon to form?

6. What indications were there that sediments were being eroded?

7. What other canyons have you heard of? How were they formed?

8. What are you wondering now?

1

A delta is a kind of landform. Deltas are made where a river flows into a large body of water. Often, it is an ocean.

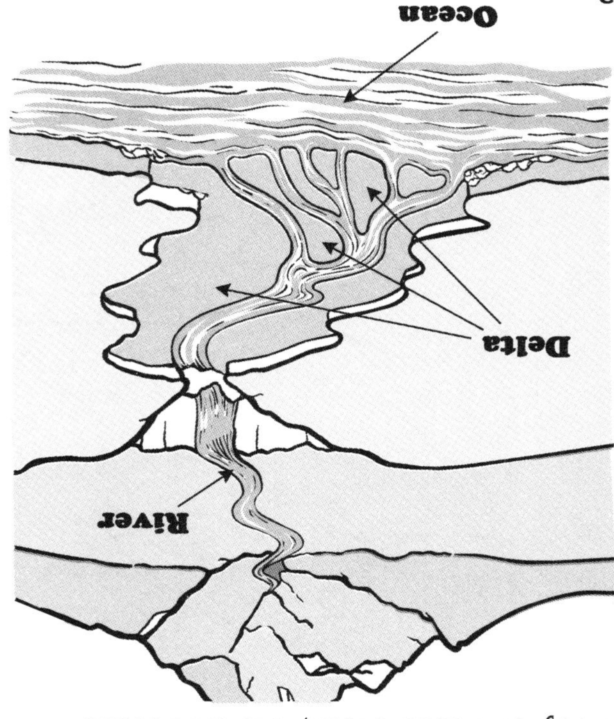


2

The shape of a delta is the reason for its name. "Delta" is actually the fourth letter of the Greek alphabet. The capital letter is shaped like a triangle (Δ). The Greek historian Herodotus first used the word "delta" to describe the Nile River delta because of its shape.

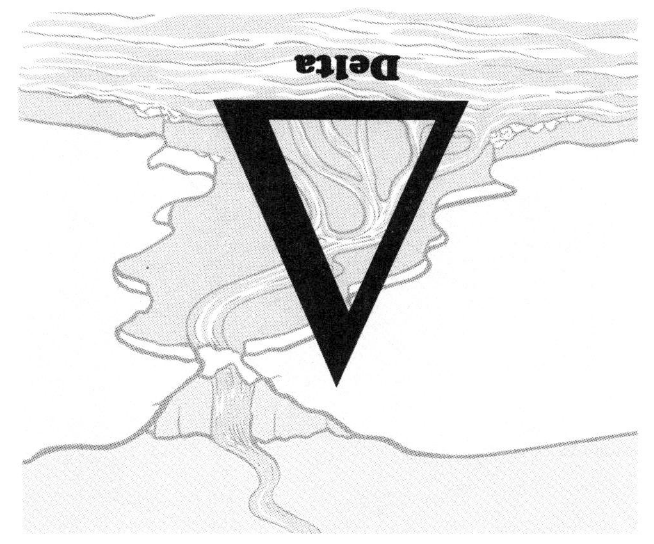


7

Deltas shift and change over time. The movements of waves, tides, and the river water can alter how they look. They are one example of how the Earth's surface is changed by water.

8

As rivers flow, they pick up sand, dirt, small rocks, bits of plants, and so on. They carry these sediments in their water as they flow downstream.

3

4

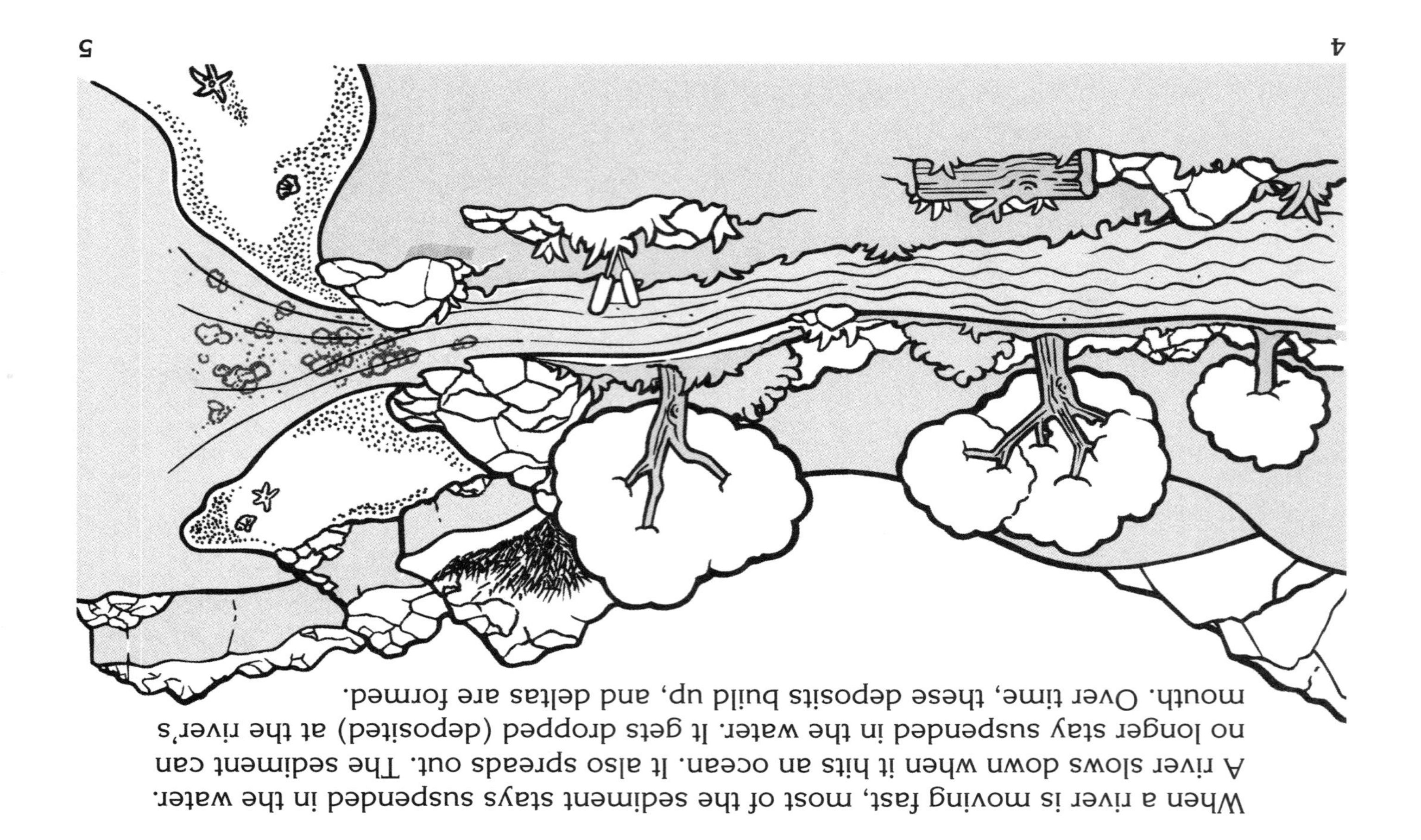

When a river is moving fast, most of the sediment stays suspended in the water. A river slows down when it hits an ocean. It also spreads out. The sediment can no longer stay suspended in the water. It gets dropped (deposited) at the river's mouth. Over time, these deposits build up, and deltas are formed.

5

When deltas form, they often have a triangular shape. This is how the Nile River delta in Egypt looks from space. Can you see the triangle?

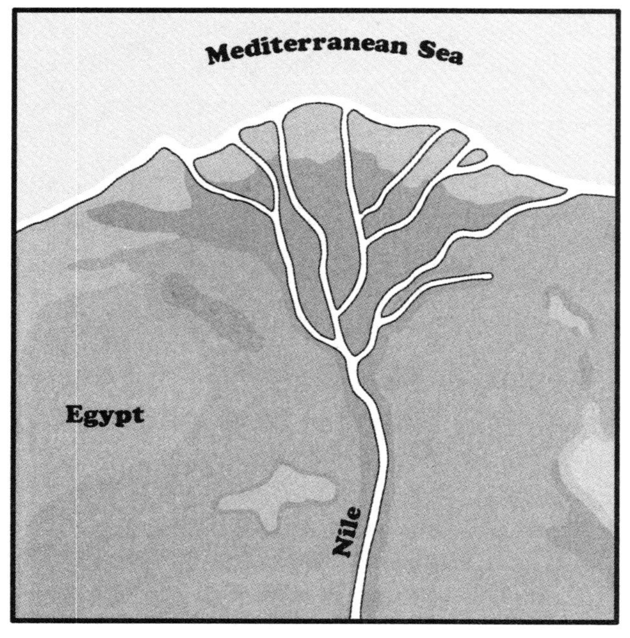


6

# A WASH OUT

**Topic**
Erosion

**Key Question**
How are plants and animals affected by erosion?

**Learning Goals**
Students will:
- build a streambed to model what happens to the land after a fire, and
- observe the effects of erosion on a tree and a muskrat's home.

**Guiding Documents**
*Project 2061 Benchmarks*
- *Changes in an organism's habitat are sometimes beneficial to it and sometimes harmful.*
- *Waves, wind, water, and ice shape and reshape the earth's land surface by eroding rock and soil in some areas and depositing them in other areas, sometimes in seasonal layers.*
- *Although weathered rock is the basic component of soil, the composition and texture of soil and its fertility and resistance to erosion are greatly influenced by plant roots and debris, bacteria, fungi, worms, insects, rodents, and other organisms.*

*NRC Standards*
- *The surface of the earth changes. Some changes are due to slow processes, such as erosion and weathering, and some changes are due to rapid processes, such as landslides, volcanic eruptions, and earthquakes.*
- *Changes in environments can be natural or influenced by humans. Some changes are good, some are bad, and some are neither good nor bad. Pollution is a change in the environment that can influence the health, survival, or activities of organisms, including humans.*

**Science**
Earth science
  erosion
Life science
  survival

**Integrated Processes**
Observing
Comparing and contrasting
Drawing conclusions
Applying

**Materials**
*For each group:*
  disposable aluminum cake pan
  soil
  sand
  large plastic cup, 16-20 oz
  Styrofoam peanut
  chenille stems, one per student
  book
  paper towels
  newspapers
  water (see *Management 7)*

*For each student:*
  student page

**Background Information**
Erosion is the carrying away of land surface by agents such as water, ice, and wind. Water is the most erosive force on Earth. Poor farming methods and wildfires are two conditions that expose the soil to the erosive forces. Vegetative cover is one of the best ways to control erosion. The roots of plants anchor the soil. The crowns of plants soften the impact of precipitation.

Erosion not only destroys the habitats of terrestrial organisms, but it also has deleterious effects on aquatic organisms. Turbidity is a measure of the amount of suspended sediment in water. When the river is carrying a heavy load of sediment, it will be the color of that sediment. There is a reduction in the amount of light that penetrates the water; this reduces the plant growth. The plants are a major source of food for many aquatic animals; therefore, their food supply is also diminished.

In this experience, students will see that the roots of a tree and the den of a muskrat can be exposed when there is a lack of protective vegetative cover.

 

**Management**

1. Students should work together in groups of four to six.
2. If working in groups of four, each group should make one tree out of the chenille stems. If working in groups of six, each group should make two trees. Cut the chenille stems in half. Each student will get both halves.

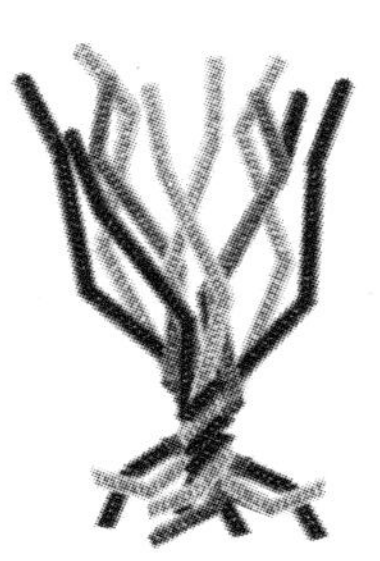

3. Break the Styrofoam peanuts into small pieces about the size of a pencil eraser. These will represent muskrats.
4. Provide groups with newspapers to absorb any spilled water.
5. Eight- or nine-inch square pans are ideal. They should be at least 1.5 inches tall.
6. Gallon-size zipper-type bags can be filled with soil and sand, making it easy for groups to carry to their work areas.
7. Use a gallon jug or 2-L bottle to dispense water to groups when ready.
8. It is assumed that students have studied erosion.

**Procedure**

1. Ask the *Key Question* and state the *Learning Goals.*
2. Have the students get into groups. Distribute the newspapers, telling students to cover their work areas.
3. Hand out the student pages and go over the construction directions.
4. Show students where the supplies are. Give them time to make their models.
5. When all groups have their models completed, go from group to group filling their cups with water.
6. At a given cue, have students "rain" on their models.
7. Allow time for groups to compare the damage caused by erosion.

**Connecting Learning**

1. What is erosion? [the carrying away of the soil]
2. What caused the erosion in our model? [water]
3. Why did the land in our model erode? [There were no plants to hold the soil.]
4. What happened to the tree? [Its roots were exposed.]
5. Why do you think this would be bad for the tree? [It couldn't support itself. It would fall over and die.]
6. The muskrat's den was exposed. How would this endanger the muskrat? [It wouldn't have protection from predators.]
7. What other plants and animals do you think would be affected by erosion?
8. Do you think that adding all that dirt to the water would hurt the fish in the stream? Explain.
9. What are you wondering now?

# A WASH OUT

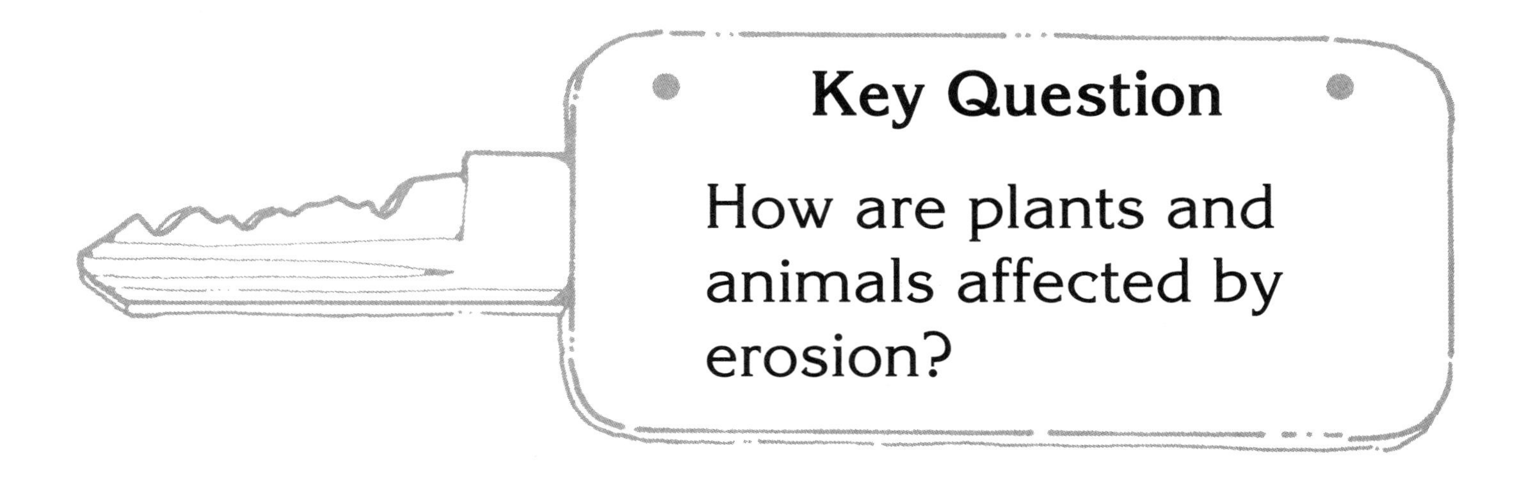


## Learning Goals

**Students will:**

- build a streambed to model what happens to the land after a fire, and
- observe the effects of erosion on a tree and a muskrat's home.

 

# A WASH OUT

**How are plants and animals affected by erosion?**

Use dirt and sand to build a streambed in the pan. Make a curvy stream. Set the pan on a book so that the pan is tilted.

Build a tree out of pipe cleaners. Make your tree have some roots.

Put your tree at the top of the pan of soil near the stream. The roots should be under the dirt.

**There is no grass because a fire has burned off the ground cover.**

Build a muskrat tunnel and den by poking your finger in the side of the banks of the stream at the top of the pan. Stick a small piece of Styrofoam in the den. This represents the muskrat.

**Now, here comes the rain.**

Fill your cup with water. Pour it out over the top of the pan. Go back and forth as you pour it out. Make sure to pour it out over the tree. Also, make sure that you pour some down the stream.

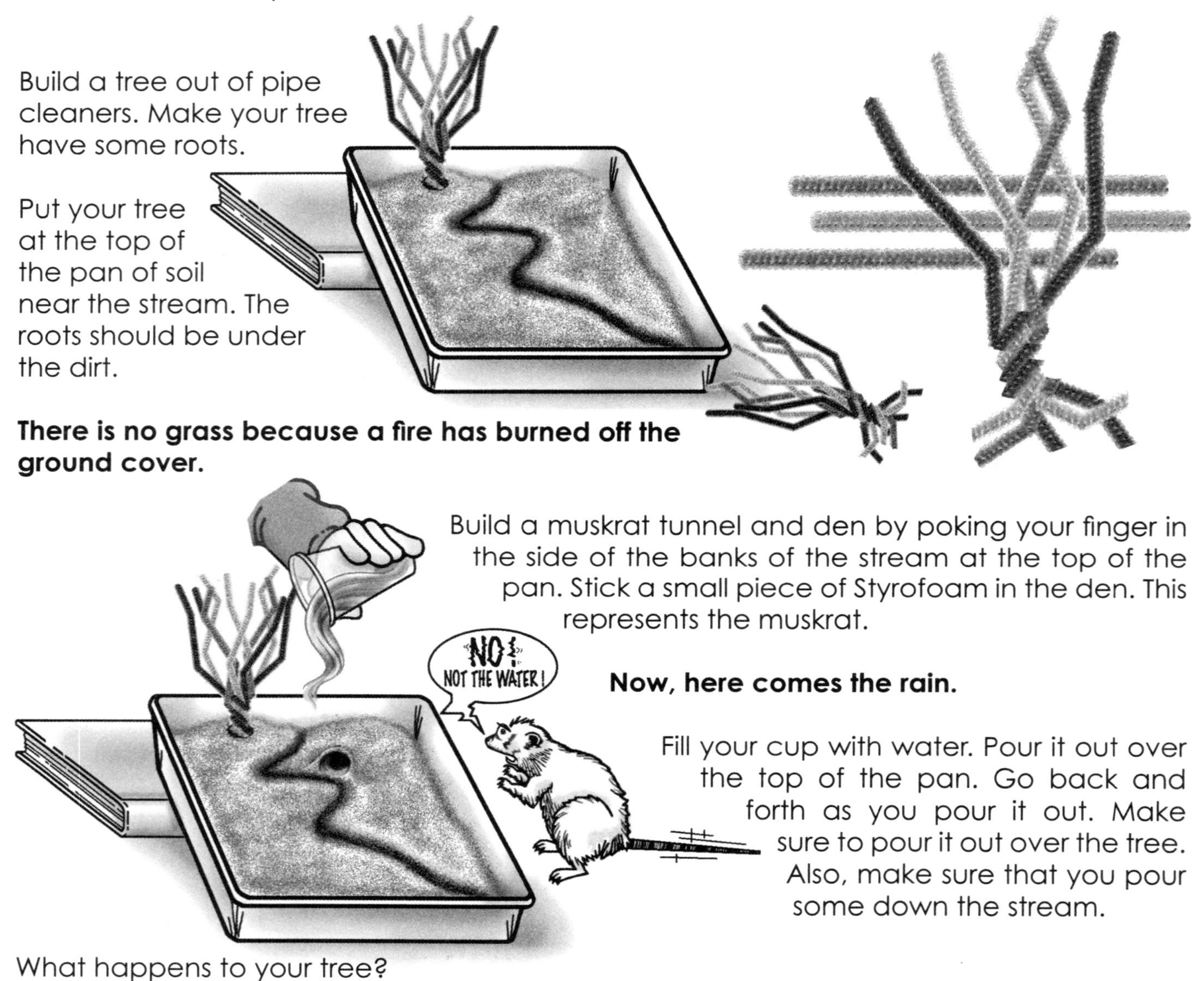


What happens to your tree?

What happens to the muskrat's home?

 

# A WASH OUT

## Connecting Learning

1. What is erosion?

2. What caused the erosion in our model?

3. Why did the land in our model erode?

4. What happened to the tree?

5. Why do you think this would be bad for the tree?

6. The muskrat's den was exposed. How would this endanger the muskrat?

 

# A WASH OUT

## Connecting Learning

7. What other plants and animals do you think would be affected by erosion?

8. Do you think that adding all that dirt to the water would hurt the fish in the stream? Explain.

9. What are you wondering now?

# Ice Sculptures

**Topic**
Glaciers

**Key Question**
How do glaciers change the surface of the Earth?

**Learning Goal**
Students will simulate how glaciers create valleys.

**Guiding Documents**
*Project 2061 Benchmarks*
- *Waves, wind, water, and ice shape and reshape the earth's land surface by eroding rock and soil in some areas and depositing them in other areas, sometimes in seasonal layers.*
- *People can often learn about things around them by just observing those things carefully, but sometimes they can learn more by doing something to the things and noting what happens.*
- *Change is something that happens to many things.*
- *A model of something is different from the real thing but can be used to learn something about the real thing.*

*NRC Standard*
- *The surface of the earth changes. Some changes are due to slow processes, such as erosion and weathering, and some changes are due to rapid processes, such as landslides, volcanic eruptions, and earthquakes.*

**Science**
Earth science
- weathering and erosion
- landforms
  - valleys
- glaciers

**Integrated Processes**
Observing
Comparing and contrasting
Identifying
Modeling

**Materials**
*For each group:*
- water
- mixing container (see *Management 2)*
- plastic spoon
- disposable pie plate
- plastic cup, 9 oz
- paper cup, 3 oz
- permanent marker
- paper towels
- student pages

*For the class:*
- plaster of Paris
- aquarium gravel
- newspaper
- access to a freezer

*For each student:*
- *Glaciers* rubber band book
- rubber band, #19

**Background Information**
Glaciers are large bodies of ice that form from accumulations of snow that have compacted over decades or centuries. Glaciers are one of the agents of erosion that have had a significant impact on the appearance of the Earth's surface. As glaciers move across the land, they erode the terrain through abrasion and plucking. Plucking is what happens when rocks from the surface of the Earth are "plucked" off and become part of the glacier. Small, embedded rock fragments in the ice cause abrasion—the glacier acts like sandpaper, smoothing and polishing the surface over which it moves. Large fragments will leave long, deep scratches as they cross over the landscape. Geologists can trace the movement of a glacier by looking at these scratches.

As glaciers move downhill, they create valleys as they carve through rocks and make hills (called moraines or drumlins) with the materials they push ahead of them or deposit after they retreat. The focus of this activity is on valleys. Glacial valleys can be identified by their characteristic "U" shape. Valleys formed by rivers have a "V" shape, but glaciers deepen and widen these valleys while smoothing them at the same time. Yosemite valley in California is a well-known example of a glacial valley.

 

**Management**

1. This activity is written as a two-day experience with students making the "glaciers" and "land" on the first day and then modeling the formation of glacial valleys on the second day. If desired, you can prepare the glaciers and land yourself and make it a one-day experience.
2. Mixing containers can be large cups (16-18 oz), liter boxes, mixing bowls, etc. They need to be able to hold about 1½ cups of plaster mixture.
3. Be sure to look at the instructions on the box of plaster of Paris for the correct water to plaster ratio. Often it is two parts plaster to one part water. Students can make an eyeball judgment for half a cup of water.
4. If you do not have a sink in your room, you will need to provide each group with a container of water. A 20-oz bottle is sufficient.
5. Have the container of plaster of Paris and the aquarium gravel in a central location where students can come and take the amount they need.
6. Having groups cover their work areas with newspaper will facilitate cleanup.

**Procedure**

*Day One: Making the Glaciers and Land*

1. Distribute the rubber band book to each student and read through it as a class. Discuss the new information students learned about glaciers and how they change the Earth's surface.
2. Tell students that they will be modeling how glaciers erode the surface of the Earth as they move, creating valleys.
3. Distribute the first student page and the necessary materials to each group. Have each group cover its work area with newspaper. Provide assistance as necessary as students make their "glaciers" and "land."
4. Provide a place where students can leave their pie plates overnight and take the glaciers to a freezer.
5. Have students clean their work areas.

*Day Two: Modeling Glacial Valleys*

1. Retrieve the cups from the freezer and distribute them to the groups along with their pie plates and paper towels. Again have them cover their work areas with newspaper.
2. Distribute the second student page and read through the instructions as a class. Allow time for groups to use their glaciers to model changing the surface of the Earth. (As the ice begins to melt and the rocks become more exposed, students will get better results as they move their glaciers over the land.)
3. When groups are done, have them place their glaciers in a large cup or bowl so that you can collect the aquarium gravel once the ice melts. Instruct them to clean their work areas.
4. Discuss what students observed as they moved their model glaciers across the plaster of Paris land. Compare this experience to glacial erosion in the real world.

**Connecting Learning**

*Day One*

1. What is a glacier? [a large body of ice]
2. How do glaciers change the Earth's surface? [They erode the Earth as they move. They pick up rocks and move them. Smaller rocks in the glacier smooth and polish rocks on the Earth's surface as they move over the top of them. Large rocks scratch the surface of the Earth. They create U-shaped valleys.]
3. Why did we put rocks in our cups of water? [to simulate the rocks that glaciers "pluck" from the ground as they move]

*Day Two*

1. What happened to the model land as you moved your glacier across it? What happened to the glacier?
2. What did you notice as the ice on the glacier began to melt? [The rocks became more exposed. They may have made deeper cuts in the plaster. Some rocks may even have fallen out of the glacier.]
3. Was it hard to push the glacier across the land? Explain.
4. How is what we did in this activity like what happens with glaciers in the real world? [Glaciers are ice and they have rocks embedded in them, just like our ice cubes. They move across the surface of the Earth, just like we moved them across our pie plates. They create valleys, and our model created valleys. Etc.] How is it different? [Real glaciers are much larger and heavier. They move more slowly. Rocks on the surface of the Earth are harder than the plaster of Paris. Etc.]
5. What landforms did your model glacier create? [valleys]
6. What do you think would happen if you left your model glacier on the land? [It would melt and leave the rocks behind.]
7. What are you wondering now?

**Extensions**

1. Explore other landforms created by glaciers, such as moraines, drumlins, fjords, and horns.
2. Freeze much larger blocks of ice and rocks and move them over a surface outside to compare the effects.

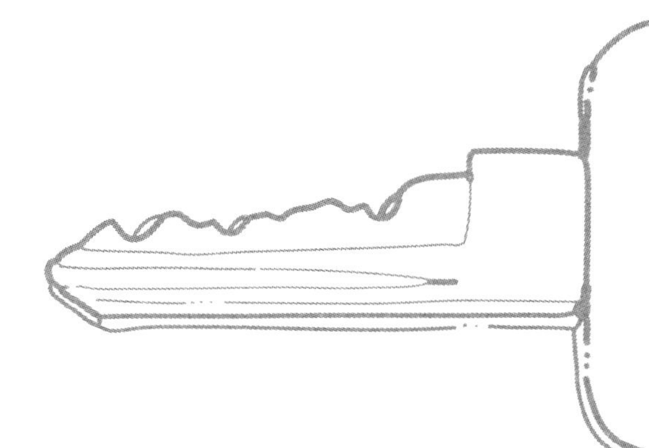

## Key Question

How do glaciers change the surface of the Earth?

## Learning Goal

**Students will:**

simulate how glaciers create valleys.

# Glaciers

Glaciers are large bodies of ice. The ice starts as snow. As more and more snow falls, it begins to compact. Over many years, the snow becomes ice.

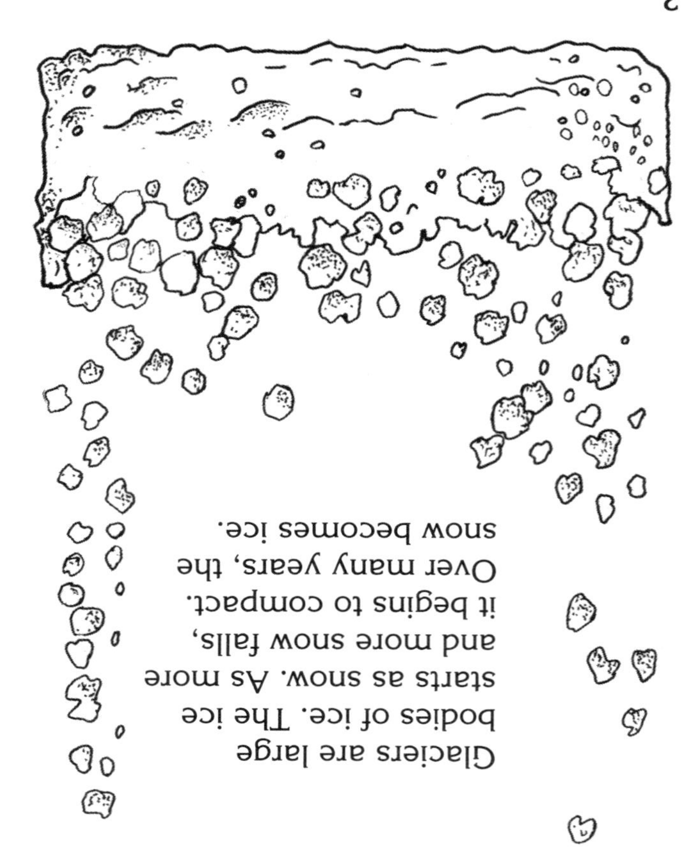



Yosemite Valley in California is a famous example of a U-shaped valley that was made by a glacier. Other U-shaped valleys can be found in Glacier National Park and in the Rocky Mountains.

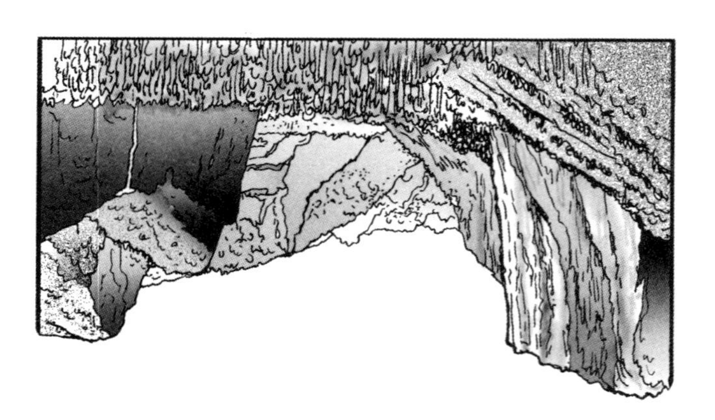

Glaciers form in the polar regions and at high elevations. As they move downhill, they erode the Earth's surface.

3

One way they cause erosion is by picking up rocks as they move. This is called "plucking." The rocks become part of the glacier until the glacier melts and retreats. Then they are left behind.

4

Another way they cause erosion is by abrasion. Think of it like sandpaper on wood. The rocks the glaciers pick up act like the grit on sandpaper. Large rocks leave scratches, like coarse sandpaper.

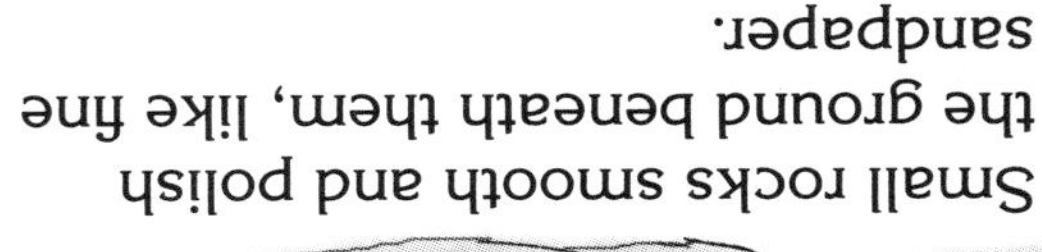

Small rocks smooth and polish the ground beneath them, like fine sandpaper.

5

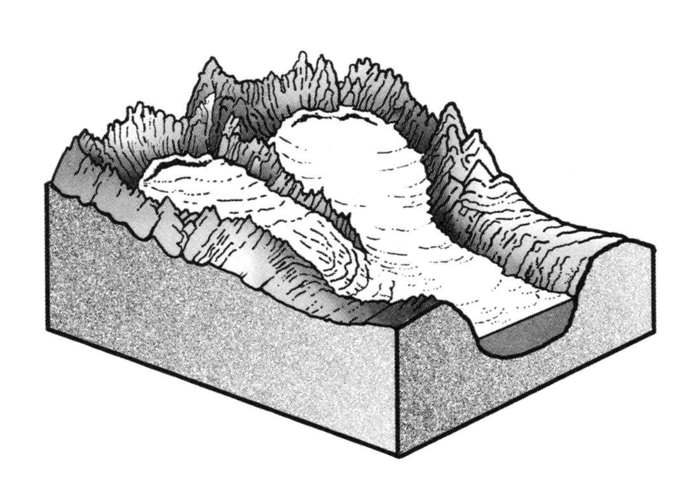

As glaciers move and smooth rocks, they form valleys. Valleys formed by glaciers have a "U" shape. This is different from valleys formed by rivers, which have a "V" shape.

6

## Make a Model Glacier

**Materials**

Aquarium gravel
Permanent marker
Paper cup, 3 oz
Water
Freezer

**Procedure**

1. Write your names on the cup.

2. Place about 1 cm of aquarium gravel in the bottom of the cup.

3. Fill the cup about ¾-full of water.
4. Place it in a freezer overnight.

## Make Model Land

**Materials**

Disposable pie plate
Water
Mixing container
Plaster of Paris
Plastic cup, 9 oz
Plastic spoon
Permanent marker

**Procedure**

1. Write your names on the bottom of the pie plate.
2. Fill the plastic cup with plaster of Paris and put it in the mixing container. Your teacher will tell you how much water to add.
3. Mix the water and plaster of Paris well. Pour the mixture into the pie plate. Try to scrape all of the plaster out of the mixing container.
4. Shake the pie plate gently from side to side to even out the surface of the plaster.
5. Allow it to harden overnight.

1. Peel the paper off your glacier. Be sure to get all of the paper off. This may be easier once it starts to melt a little.

2. Hold the glacier with a paper towel to protect your hand. Hold it so that the rocks are on the bottom.

3. Take turns in your group moving the glacier over your land. Remember that glaciers are huge and very heavy, so push as hard as you can.

4. Observe the land and the bottom of the glacier. Record what you observe. Illustrate how your land looks after all group members have had a chance to move the glacier.

5. How does this experience model the way real glaciers change the surface of the Earth? What landform(s) did your model glacier create?

*Day One*

1. What is a glacier?

2. How do glaciers change the Earth's surface?

3. Why did we put rocks in our cups of water?

*Day Two*

1. What happened to the model land as you moved your glacier across it? What happened to the glacier?

2. What did you notice as the ice on the glacier began to melt?

3. Was it hard to push the glacier across the land? Explain.

## Connecting Learning

4. How is what we did in this activity like what happens with glaciers in the real world? How is it different?

5. What landforms did your model glacier create?

6. What do you think would happen if you left your model glacier on the land?

7. What are you wondering now?

**Topic**
Wind erosion

**Key Question**
What causes the sand and/or snow to form drifts?

**Learning Goals**
Students will:
- explore the effects of wind as an agent of erosion in the picking up of sand and snow, and
- note how sand and snow are deposited when they encounter obstacles.

**Guiding Documents**
*Project 2061 Benchmarks*
- *Waves, wind, water, and ice shape and reshape the earth's land surface by eroding rock and soil in some areas and depositing them in other areas, sometimes in seasonal layers.*
- *People can often learn about things around them by just observing those things carefully, but sometimes they can learn more by doing something to the things and noting what happens.*
- *Change is something that happens to many things.*
- *A model of something is different from the real thing but can be used to learn something about the real thing.*

*NRC Standard*
- *The surface of the earth changes. Some changes are due to slow processes, such as erosion and weathering, and some changes are due to rapid processes, such as landslides, volcanic eruptions, and earthquakes.*

**Science**
Earth science
 erosion

**Integrated Processes**
Observing
Comparing and contrasting
Collecting and recording data
Interpreting data
Drawing conclusions

**Materials**
Fine sand (see *Management 4)*
Shoebox lid
Drinking straws
Newspapers
Centicubes
Twigs
Pieces of grass, straw, or other vegetation
Pebbles
Hand-held hair dryer
Safety goggles

**Background Information**
The Earth is a dynamic body; it is always changing. The natural processes that include tectonic and volcanic activities are elevating parts of the Earth's surface. Opposing processes of weathering, erosion, and mass wasting are continually moving materials from higher elevations to lower elevations. This activity deals with the process of *erosion,* which is defined as a process in which sediment is loosened and moved from one place to another. Agents of erosion include water, wind, ice, and gravity.

Winds do very little in the way of eroding large masses of solid rock other than to act as a sandblasting agent. However, when physical or chemical weathering fragments the rock, small particles can be lifted by the wind. Besides particle size, moisture is another factor that affects wind erosion. Small particles of fragmented rock and soils are held together by moisture; thus, winds erode most effectively in arid climates where the winds tend to be dry and moisture quickly evaporates. (An exception to this is the dune belt along the ocean coasts or large lakes where the sand dries so quickly in the wind that it moves even in humid climates. Notice, however, that in these areas the dunes become covered with soil and vegetation not too far inland and sand no longer blows.)

Wind is a transport agent of sand. The sand moves in a combination of slides and rolls along the surface and by saltation. *Saltation* is the temporary suspension of sand grains in the air. Sand grains are lifted into the air then fall back to the ground where they strike other grains lying on the surface with enough force to put those grains into motion. The result of these movements of sand is the formation of sand ripples and dunes that migrate in the direction of the wind.

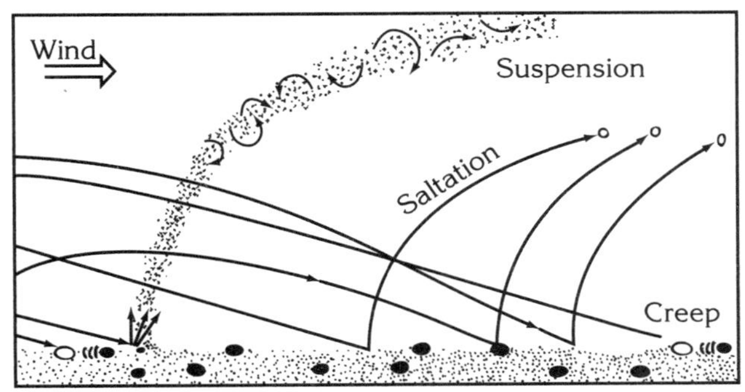


 

Sand dunes are found behind beaches along large lakes or ocean coasts, in the sandy floodplains of large rivers in arid and semiarid regions, and in desert regions. All these places have a large supply of loose sand and sufficient wind power. The major ingredients for a sand dune are ample sand, ample wind, and an obstacle such as a rock or clump of vegetation.

The streamlines of wind separate around the obstacles and rejoin downwind. This creates a wind-shadow downwind of the obstacle. In the wind-shadow zone, the wind velocity is much lower than in the main flow. This lower velocity allows the sand grains to settle and accumulate without being picked up and redistributed.

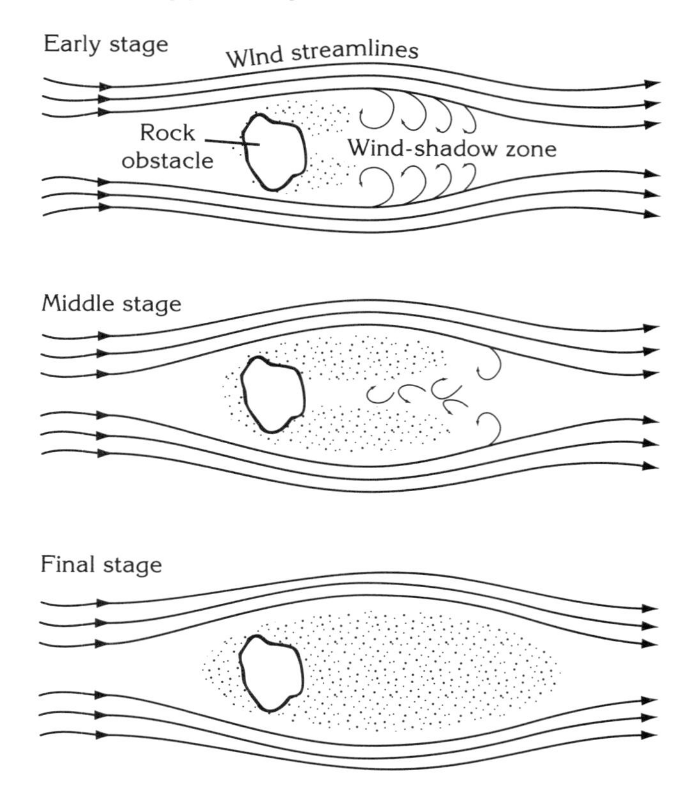


As this process continues, the sand dune itself becomes an obstacle. Under the right conditions, it continues to grow and starts to migrate downwind as particles are blown on the windward side of the dune.

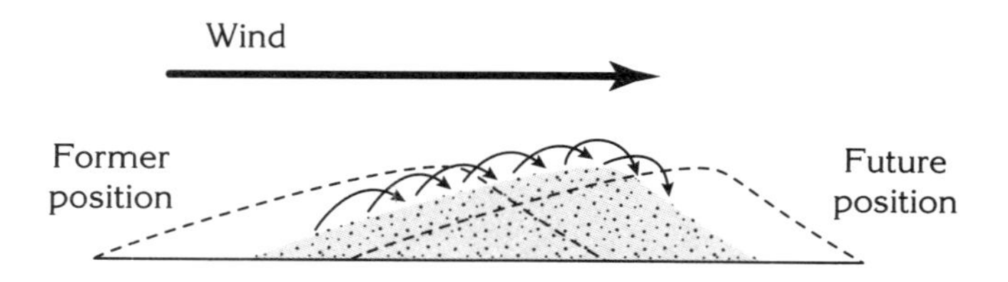


In this activity, students will explore the effects of obstacles on the drifting of sand. Similar characteristics can be seen in the drifting of snow. Although snow involves many variables that may not apply to sand, students living in cold, snowy regions will be able to observe the similarities of the effects of obstacles on drifting.

**Management**

1. This activity can be set up at a center or done by students in groups of four or five.
2. *Sand Dunes and Snow Drifts* is written in a very open-ended format. Students will need to have ample time or many revisitations to the activity in order to explore various arrangements of obstacles. It is suggested that students record the placement of their obstacles and an illustration of the drifts that form and post these for observation by other students.
3. You will need to use the low/cool speed on the hair dryer.
4. If fine sand is not available, salt can be substituted.
5. Precautions need to be taken because of the blowing sand. It is recommended that students wear safety goggles and stand behind the person holding the hair dryer. The area where students are working should be lined with newspaper to aid in the clean up. Warn students to take care if any sand accumulates on the floor because the floor may become very slick.
6. Ask students to bring in shoebox lids.
7. Centicubes (item number 1952) are available from AIMS.

**Procedure**

1. Ask the *Key Question* and state the *Learning Goals.*
2. Have students fill their shoebox lids with sand. Go over safety procedures!
3. Allow them time to create an outdoor setting using centicubes or other blocks for buildings and twigs, pebbles, etc., for landscaping. They may want to form small hills and valley areas. The fewer the obstacles at first, the better the generalizations of their effects.
4. Ask students to predict what they think will happen when they hold a hair dryer parallel to the surface of the "sand or snow" and turn it on.
5. Have students use the low setting on the dryer and hold it at surface level about 10 cm from the front end of the model.
6. After the students have observed the movement of the sand, have them record their obstacle placement and results of wind erosion on the student page.
7. Direct them to move the placement of their obstacles and blow the sand again. Allow students to investigate in this manner as long as they are actively involved in the learning process.
8. Have students build sand fences (snow fences) by cutting drinking straws into four centimeter pieces and placing the pieces upright in the sand. Tell them to place their fences in areas where they want to control the drifting of the sand.

 

## Connecting Learning

1. What is erosion? How did this lesson model erosion?
2. What did you notice about the effects of the wind on the windward side (the front side) of the obstacles? Were these effects seen with all obstacles? Explain.
3. What happened in areas directly behind the obstacles? Why?
4. What happened in areas where an obstacle was downwind of another obstacle? Give reasons why you think this happened.
5. Can you think of any instances in which you would want a drift to form in a certain area? If so, explain that instance and what you would need to do to accomplish this.
6. Road departments are often faced with sand/snow drifts that hinder traffic from moving. What suggestions would you give them for preventing this from happening?
7. What would happen if the wind blew from a different direction?
8. Explain in your own words any new things you learned in this investigation.
9. What are you wondering now?

## Extensions

1. Have students do research on the different types of sand dunes and what causes them.
2. Locate and assemble a montage of pictures of sand dunes and snow drifts.
3. Have students investigate droughts and their effects on soil erosion. Direct them specifically to the Dust Bowl of the 1930s.

# Sand Dunes and Snow Drifts

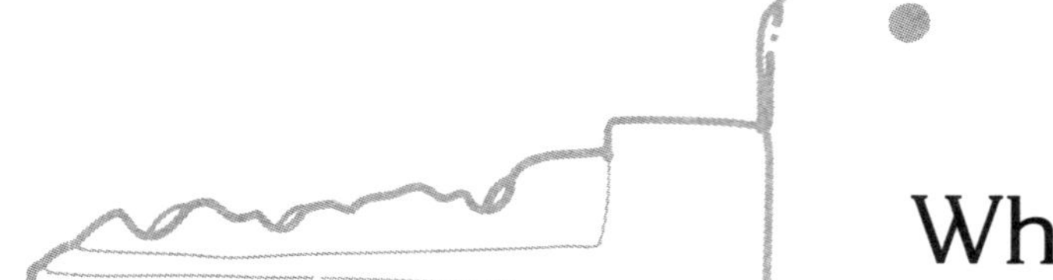

## Key Question

What causes the sand and/or snow to form drifts?

## Learning Goals

**Students will:**

- explore the effects of wind as an agent of erosion in the picking up of sand and snow, and
- note how sand and snow are deposited when they encounter obstacles.

# Sand Dunes and Snow Drifts

In your journal, record at least three observations you made about wind erosion.

CONNECTING LEARNING

## Connecting Learning

1. What is erosion? How did this lesson model erosion?
2. What did you notice about the effects of the wind on the windward side (the front side) of the obstacles? Were these effects seen with all obstacles? Explain.
3. What happened in areas directly behind the obstacles? Why?
4. What happened in areas where an obstacle was downwind of another obstacle? Give reasons why you think this happened.

## Connecting Learning

5. Can you think of any instances in which you would want a drift to form in a certain area? If so, explain that instance and what you would need to do to accomplish this.

6. Road departments are often faced with sand/snow drifts that hinder traffic from moving. What suggestions would you give them for preventing this from happening?

7. What would happen if the wind blew from a different direction?

CONNECTING LEARNING

## Connecting Learning

8. Explain in your own words any new things you learned in this investigation.

9. What are you wondering now?

**Topic**
Erosion and deposition

**Key Question**
Where can we find examples of erosion and deposition on our school grounds?

**Learning Goal**
Students will look for examples of erosion and deposition on the school grounds.

**Guiding Documents**
*Project 2061 Benchmarks*
- *Waves, wind, water, and ice shape and reshape the earth's land surface by eroding rock and soil in some areas and depositing them in other areas, sometimes in seasonal layers.*
- *Change is something that happens to many things.*

*NRC Standard*
- *The surface of the earth changes. Some changes are due to slow processes, such as erosion and weathering, and some changes are due to rapid processes, such as landslides, volcanic eruptions, and earthquakes.*

**Science**
Earth science
  erosion
  deposition

**Integrated Processes**
Observing
Collecting and recording data
Comparing and contrasting
Predicting
Applying

**Background Information**
Weathering, erosion, and deposition are three of the processes that change the Earth's landscape. Mountains and hills are gradually flattened and valleys widened into huge plains through these processes.

Weathering is a term that includes all the changes in rock materials that result from their exposure to the atmosphere. It transforms solid bedrock into small fragments that can be removed by agents of erosion.

Erosion is the moving of Earth's surface by natural forces. The means for this movement of Earth's crust can be wind, water, and ice. Whatever the method, Earth is constantly being altered. The process of erosion is often very slow and difficult to observe, but over thousands of years, the mountains and hills have been worn down and rivers have widened their valleys into broad plains.

Water is by far the most powerful agent of erosion. More than one-quarter of the annual precipitation falling onto the continents runs off into the ocean via rivers, streams, etc. Rivers carry away rocks and soils, eroding the mountains and hills and carving out valleys and canyons. Streams of water roll materials downhill onto the lowlands or out to sea. Such erosion is often easily seen in coastal areas and along rivers and streams where noticeable amounts of land can be lost each year. Water continuously erodes and changes Earth's surface.

Wind erosion occurs mostly along the ground surface. Wind carries off soil and small rocks. The surfaces of boulders and rocks may eventually be pitted or worn smooth by the flying dust and sand grains. During the 1930s, large amounts of topsoil were lost when areas of the middle west received little rain. Plants died and could no longer anchor the soil when winds blew. This area and period of time became known as the Dust Bowl.

Snow collects in hollows of mountains. As the weight of the snow builds up, it becomes compressed and forms ice. As more snow is added and turns to ice, gravity and the weight of the ice cause it to slowly move downhill; thus a glacier is formed. Glaciers carry embedded rocks and soil a great distance before dropping them. The glaciers also act like bulldozers by pushing rocks in front of them. Rocks stuck in the bottom and sides of glaciers scrape, scratch, and dig into the rocks beneath the glacier. Glaciers can also cause erosion as the ice melts and slowly flows downhill.

Deposition is the laying down of eroded materials. It is evident in the bends of rivers and the deltas that form as major rivers drain into the oceans. Materials deposited by winds are found in the lee of objects and structures. Glacial deposits result as the glaciers melt leaving ridge-like hills.

Although erosion is a natural process, people can increase the effect by clearing land of vegetation or improper cultivation. They can also slow down the process by planting cover crops on bare land, terracing land, building wind breaks, etc.

**Management**

1. Look around the school ground or neighborhood for places where wind, water, or people have eroded the area and for areas where deposition has occurred. For real-world examples of water erosion, if there are no gutters or drain spouts, simply pour a bucket of water over some loosely packed soil so that students can get an idea of what happens with that process.
2. Select some objects in the classroom that students can pick up, carry, and deposit elsewhere to show change. Some suggested items are chairs, books, plants, etc.

**Procedure**

1. Introduce the lesson and the key terms by doing some class role-playing. Tell the students that they are going to change the room a bit. Inform them that the student doing the change represents erosion. Tell them that the job of erosion is to pick up, carry, and deposit (put things down) items from around the room. Inform them that for the first few times, you will be the director and will tell the student what to pick up, where to carry it, and where to deposit it.
2. Choose a student to represent erosion. Direct him/her to pick up, carry, and deposit some items from around the room. Have different students repeat this procedure. Emphasize the processes and what changes resulted in the room from the "erosion" and "deposition." Select a student to be director and continue the role-play.
3. Take the students outside to look at a doormat. Ask them what they see on or under the mat. Ask them how the dirt (grass, sand, etc.) got there. Ask them if they think erosion and deposition have been at work. [This is a case of pick up, carry, and deposit. The soil, etc., was picked up on their shoes, carried to the area, and deposited when they scraped their feet. They caused the erosion and resulting deposition.]
4. Tell the students that in nature, the processes of erosion and deposition change the way Earth looks. Ask them what they think does the picking up and carrying. [wind, water, ice, people]
5. If it is possible, find areas on the school ground where erosion can be shown—gullies formed by running water, hard-packed paths formed by human feet, or blowing sand and dirt. Also look for locations where deposition has taken place—piles of dust in the corners of buildings, dirt and sand collecting around doormats, or mounds of soil in the corners of flowerbeds.
6. Back in the classroom, list the agents of erosion [wind, water, ice, people] on the board. Also list places where the students have seen erosion and deposition.
7. Compare the examples of erosion and deposition they saw on the playground to larger-scale examples that would be found elsewhere.

**Connecting Learning**

1. In the first part of the activity, how did you change the room?
2. How is this similar to what happens in nature? [Natural forces such as water, wind, and ice erode the surface of the Earth and carry eroded particles away from where they were. The eroded particles are deposited in other places.]
3. What were some places where you saw erosion on the school grounds?
4. What were some places where you saw deposition on the school grounds?
5. How do these examples you saw compare to what you might see in other places? [We were looking at small examples of erosion and deposition. These processes also take place on a much larger scale. Rivers erode rocks and carry sand and sediment and deposit it in deltas. Glaciers carve out canyons and move tons of rock at a time. Etc.]
6. What are you wondering now?

# Confirming Changes

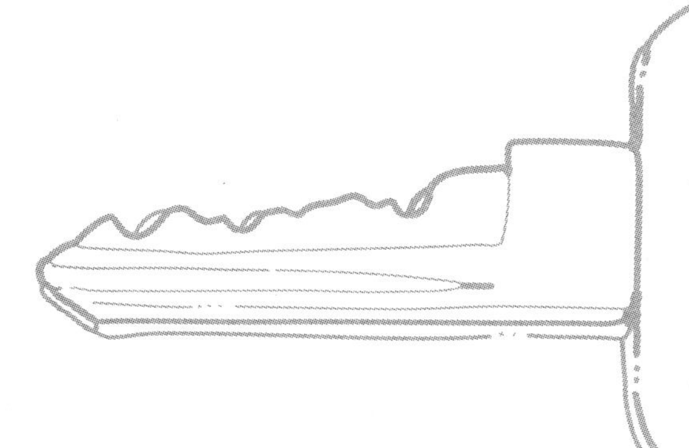

## Key Question

Where can we find examples of erosion and deposition on our school grounds?

## Learning Goal

look for examples of erosion and deposition on the school grounds.

1. In the first part of the activity, how did you change the room?

2. How is this similar to what happens in nature?

3. What were some places where you saw erosion on the school grounds?

4. What were some places where you saw deposition on the school grounds?

5. How do these examples you saw compare to what you might see in other places?

6. What are you wondering now?

**Topic**
Erosion

**Key Question**
How do wind, water, and ice change rocks and soil over time?

**Learning Goals**
Students will:
- identify evidence of erosion,
- create models depicting various types of erosion,
- explain the causes of erosion, and
- research and share information about changes that have occurred over time due to erosion.

**Guiding Documents**
*Project 2061 Benchmarks*
- *Waves, wind, water, and ice shape and reshape the earth's land surface by eroding rock and soil in some areas and depositing them in other areas, sometimes in seasonal layers.*
- *People can often learn about things around them by just observing those things carefully, but sometimes they can learn more by doing something to the things and noting what happens.*
- *Change is something that happens to many things.*
- *A model of something is different from the real thing but can be used to learn something about the real thing.*

*NRC Standard*
- *The surface of the earth changes. Some changes are due to slow processes, such as erosion and weathering, and some changes are due to rapid processes, such as landslides, volcanic eruptions, and earthquakes.*

**Science**
Earth science
  erosion

**Integrated Processes**
Observing
Collecting and recording data
Comparing and contrasting
Predicting
Applying

**Materials**
Potted plant
Soil
Sand
Rocks
Disposable aluminum trays
Book
Clay
Sugar cubes
Cup
Newspapers
Ice (see *Management 2)*
Safety goggles
Containers for water (see *Management 4)*
Paper towels
Zipper-type plastic bags, quart size
Plastic bowls
Paper cups, 5 oz
Toothpicks
Metric rulers
Digital camera
*Evidence for Erosion* journal
Canyon and glacier photos, included

**Background Information**
*Erosion* is the picking up and carrying of Earth's materials to a different place. The agents of erosion are wind, water, and ice. Whatever the method, Earth is constantly being changed. The process of erosion is often very slow and not easily noticed, but over thousands of years, the mountains and hills have been broken down by weathering and carried by wind, water, and ice down the slopes.

Water is by far the most powerful erosive agent. More than one-quarter of the annual precipitation falling onto the continents runs off into the ocean through rivers, streams, etc. Rivers carry away rocks and soils, eroding the mountains and hills and carving out valleys and canyons.

Streams of water, pulled by the force of gravity, roll materials downhill onto the lowlands or out to sea. Such erosion is often easily seen in coastal areas and along rivers and streams where noticeable amounts of land can be lost each year.

Wind erosion occurs mainly along the ground surface. Wind carries off soil and small rocks. The surfaces of boulders and rocks may be pitted or worn smooth by the flying dust and sand grains. During the 1930s, areas of the Midwest received little rain. Plants died and could no longer anchor the soil when winds blew. Poor farming practices that utilized straight furrows added to the problem. This area and period of time became known as the Dust Bowl.

 

Snow collects in hollows of mountains. As the weight of the snow builds up, it becomes compressed and forms ice. As more snow is added and turns to ice, gravity and the weight of the ice causes it to slowly move downhill. A glacier is formed. Glaciers carry embedded rocks and soil great distances before dropping them. The glaciers also act like bulldozers by pushing rocks in front of them. Rocks stuck in the bottom and sides of glaciers dig into the rocks beneath the glacier. Glaciers can also cause erosion as the ice melts and slowly flows downhill.

Although erosion is a natural process, people can increase the effect by clearing land of vegetation or improper cultivation. They can also slow down the process by planting cover crops on bare land, terracing land, building windbreaks, etc.

**Management**

1. Students should take a pencil, notebook, and digital camera outside to document evidence of erosion on the school property.
2. A large block of ice is needed for *Part Four* of this activity. This ice block can be purchased from a local ice vendor or made in the school freezer. Simply line a large plastic tub, such as an ice cream container or bucket, with plastic wrap or coat with a cooking spray, fill with water, and freeze until solid.
3. Set up stations for *Part Three*. Materials for each station are on the station cards.
4. Two-liter bottles can serve as water sources for the stations in *Part Three*. A two-liter bottle with the neck cut off can serve as the water collection container.
5. Caution students about the dangers of blowing sand. Safety goggles should be worn during the wind erosion investigation.
6. Each student will need an *Evidence for Erosion Journal* for *Part Three*. Cut the pages apart, order them, and staple along the left side.
7. You will need a computer with a projector in order to show the students the photos of the canyons and glaciers found on the CD.
8. Metric rulers (item number 1909) are available from AIMS.

**Procedure**

*Part One—Evidence of Erosion*

1. Remove a potted plant from the pot, keeping the soil intact. Discuss how the roots of the plant help hold the soil in place. Ask the students what they think would happen if the plant were in the ground instead of a pot. Ask what might happen if water kept running over the plant.
2. Introduce the term *erosion* (the process by which bits of rock and soil are picked up and carried elsewhere) and discuss how wind, water, and ice can cause erosion. Ask students if and where they have ever seen the effects of erosion.
3. Explain that the class is going to go out to the playground to examine the effects of erosion on their own playground. Ask students to remember how plants hold soil and to pay special attention to the placement of trees and shrubs on the school grounds.
4. Point out evidence of erosion on the school grounds. (Some good examples are often near drains, drainpipes, around the doormat, in the corners of the school, and at the edges of the blacktop.)
5. Set boundaries for where students can explore safely. Have them pair up to look for other signs of erosion.
6. Tell the students that when they find examples of erosion, they should describe them in their notebooks, draw labeled sketches, and take digital pictures.
7. After students are back in the room, ask them to share what they have written in their journals about the effects of erosion on the playground and school property.
8. Discuss what the students think caused the various examples of erosion on the playground. Ask if they think the trees and shrubs were placed in particular areas to help stop the effects of erosion.

*Part Two—Wind Erosion*

1. Give each pair of students a disposable aluminum-baking tray, enough soil to fill the tray, a small amount of sand, newspapers, and some rocks. Cover each working area with newspapers.
2. Instruct students to fill their tray with soil, patting it down firmly in place. Have students lightly sprinkle sand over the entire area.
3. Ask students to predict what will happen if the land experiences a gentle breeze. ...a wind storm.
4. Tell students to put on their safety goggles. Have one student from each pair gently blow across the plot of land. Ask what changes, if any, took place in their trays.
5. Direct the other student to blow harder across the plot of land to simulate a windstorm.
6. Ask students to see if the soil or sand is moving out of position.
7. Have students predict what might happen if there were rocks or other obstacles on the land.
8. Instruct students to place a few rocks randomly on the plot of land and repeat the blowing process.
9. End with a discussion about how wind erodes the land and how wind erosion can be prevented or lessened.

 

*Part Three—Water Erosion*

1. Review the physical features of the Earth's surface. [canyons, mountains, etc.]
2. Show the students photos of the Grand Canyon and Providence Canyon found on the CD. Explain how the canyons were formed by water erosion.
3. Tell the students that they are going to work through stations that deal with different types of water erosion.
4. Introduce the class to the erosion stations set up around the room. Distribute the *Evidence for Erosion Journals*.

*Part Four—Glacial Erosion*

1. Describe what a glacier is and how it moves down a mountain. Show the photos of the glaciers found on the CD. Explain that glaciers are like rivers of packed snow and ice. Tell students that they are going to model how glaciers erode Earth's materials.
2. Place a block of ice on the ground outside. Put a pad of newspaper over the ice. Invite a student to sit on the block and have another student push the block slowly across the ground. Remind students that gravity pulls glaciers slowly down the hill.
3. Every two meters have the students stop and look at the bottom of the ice block. Ask them to observe and record what the ice has picked up and what has been pushed out of the way or ahead of the block.
4. Repeat the process with different ground surfaces, such as on the grass, on the sidewalk, in a sandy area, etc.
5. Have students observe what happened to the particles that have been picked up by the ice. Discuss the changes in the surface on which the ice slid.
6. End with a discussion/review about how glaciers plow the land as they travel.
7. Show the students a map of the United States. Point out New York state. Have them notice Long Island. Tell them that millions of years ago a glacier moved down into this area. When it melted, it left the dirt and rock that it had pushed and carried. The dirt and rocks it left make up Long Island.

*Part Five—Research*

1. Have students brainstorm various landforms on the Earth's surface. (mountains, canyons, caves, etc.) Ask how they think each was made and whether they think erosion could have been involved.
2. Have students view the effects of wind, water, and ice on soil and rocks by going to these and other sites. Instruct students to read the information and view the pictures.
   - http://www.weru.ksu.edu/new_weru/multimedia/multimedia.html (dust storm/wind erosion pictures)
   - http://www.nrcs.usda.gov/technical/ECS/agronomy/photos.html (wind and water erosion pictures)
   - http://www.earthscienceworld.org/images/search/index.html (search for "erosion" to get photos)
   - http://www.uwgb.edu/DutchS/EarthSC202Slides/WINDSLID.htm (photos and descriptions of wind erosion)
   - http://www.wtamu.edu/~crobinson/Erosion/index.html (photos and descriptions of wind and water erosion)
3. Divide the class into groups of three to research unique landforms such as Carlsbad Caverns in New Mexico, Antelope Canyon in Utah, and Pumpkin Patch Concretions in California. Instruct them to find answers to the following questions:
   - Where is the landform located?
   - How did the landform develop?
   - What forces of nature caused it to form?
4. End with a time for students to share their information and have a discussion about what the students think are the most important points relating to erosion.

**Connecting Learning**

1. What is erosion?
2. What evidence of erosion did you see as you walked around the school?
3. How does water change the Earth?
4. How does wind change the Earth?
5. How does ice change the Earth?
6. What happened to the sand as you blew across your plot of land? How was the effect different when you added rocks and blew?
7. What are some examples of wind erosion in nature?
8. How is the way a cave and a canyon are formed different? How is the process similar?
9. What role does water play in shaping the Earth?
10. Is a hurricane or tornado an agent of erosion? Explain your answer. [Yes. A hurricane and tornado are violent winds. They pick up and carry dirt and rocks.]
11. What are you wondering now?

# EVIDENCE FOR EROSION

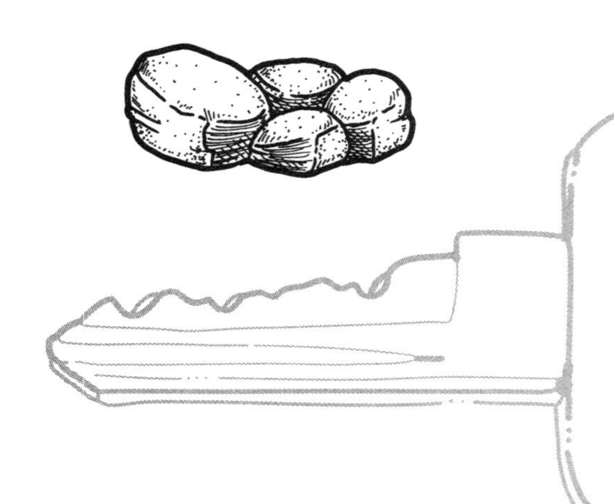

## Key Question

How do wind, water, and ice change rocks and soil over time?

## Learning Goals

**Students will:**

- identify evidence of erosion,
- create models depicting various types of erosion,
- explain the causes of erosion, and
- research and share information about changes that have occurred over time due to erosion.

Glacier 1

Glacier 2

Glacier 3

Glacier 4

Grand Canyon 1

Grand Canyon 2

Grand Canyon 3

Grand Canyon 4

Providence Canyon 1

Providence Canyon 2

Providence Canyon 3

 

## STATION 1 What happens when it rains on a sandy plain?

Materials: plastic bowls, sand, toothpick, 5-oz paper cups, water, ruler

1. Fill the small bowl with sand.
2. Smooth the sand flat so it fills the whole bowl to form a "plain."
3. Use a toothpick to make 10 small holes in the bottom of one of the paper cups.
4. Fill a second cup with water.
5. One student: Hold the cup with holes 30 centimeters above the plain.
6. Second student: Gently pour the water from the other cup into the cup with holes.
7. Watch what happens, and answer the questions in your journal.

## STATION 2 How does water erode a sandy mountain?

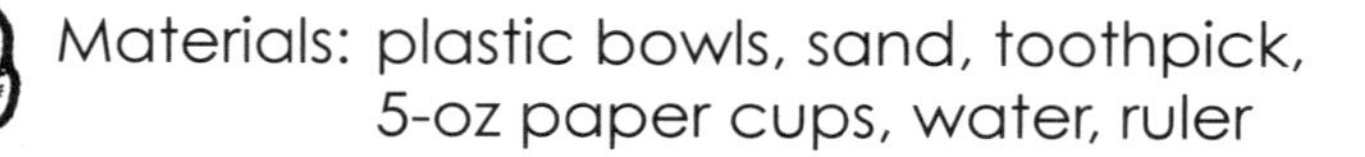

Materials: plastic bowls, sand, toothpick, 5-oz paper cups, water, ruler

1. Fill the small bowl with sand.
2. Shape a pile of sand into a "mountain."
3. Use a toothpick to make 10 small holes in the bottom of one of the paper cups.
4. Fill the second cup with water.
5. One student: Hold the cup with holes 30 centimeters above the center of the mountain.
6. Second student: Gently pour the water from the other cup into the cup with holes.
7. Watch what happens, and answer the questions in your journal.

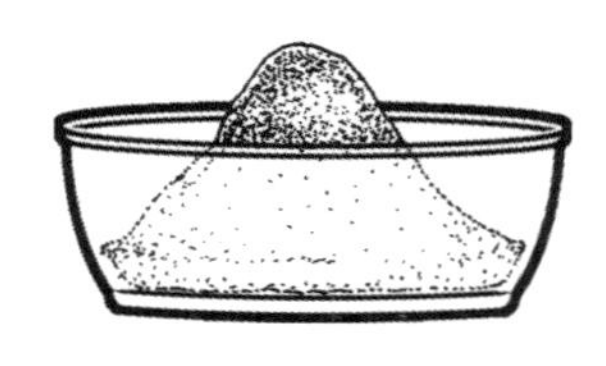

 

## STATION 3

## How does grass affect water erosion on a mountain?

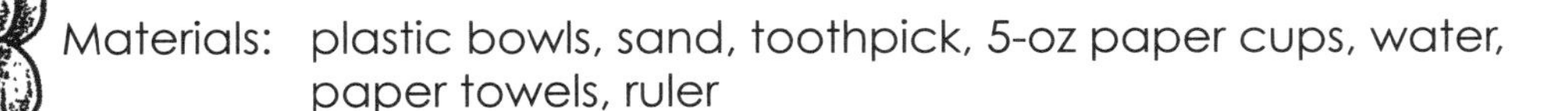

Materials: plastic bowls, sand, toothpick, 5-oz paper cups, water, paper towels, ruler

1. Fill the small bowl with sand.
2. Shape a pile of sand into a mountain.
3. Pretend to grow grass all over the mountain by covering it with a paper towel. Pat the paper towel down so that it is touching the sand everywhere.
4. Use a toothpick to make 10 small holes in the bottom of one of the paper cups.

5. Fill the second cup with water.
6. One student: Hold the cup with holes 30 centimeters above the center of the mountain.
7. Second student: Gently pour the water from the other cup into the cup with holes.
8. Watch what happens, and answer the questions in your journal.

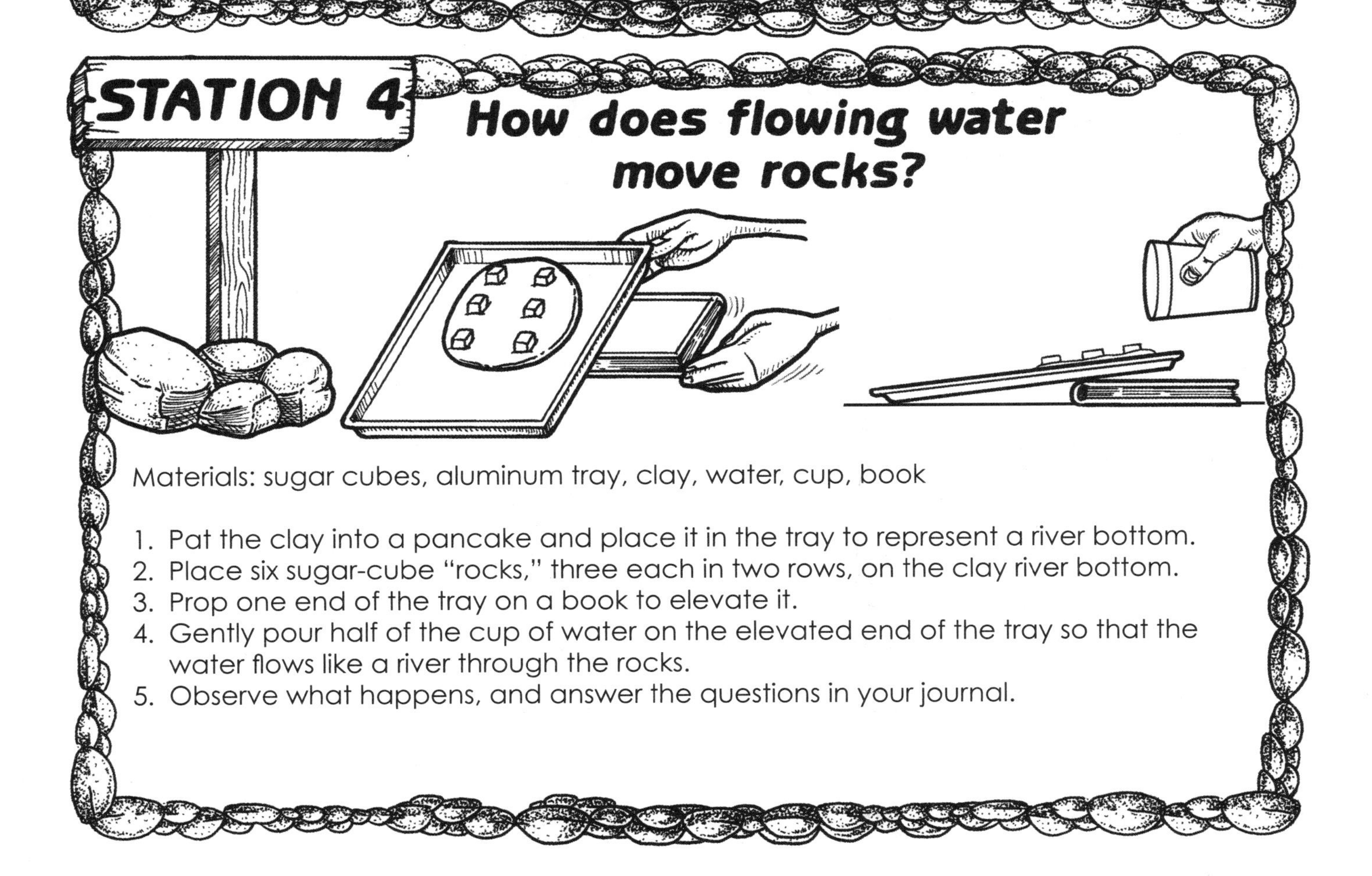

## STATION 4

## How does flowing water move rocks?

Materials: sugar cubes, aluminum tray, clay, water, cup, book

1. Pat the clay into a pancake and place it in the tray to represent a river bottom.
2. Place six sugar-cube "rocks," three each in two rows, on the clay river bottom.
3. Prop one end of the tray on a book to elevate it.
4. Gently pour half of the cup of water on the elevated end of the tray so that the water flows like a river through the rocks.
5. Observe what happens, and answer the questions in your journal.

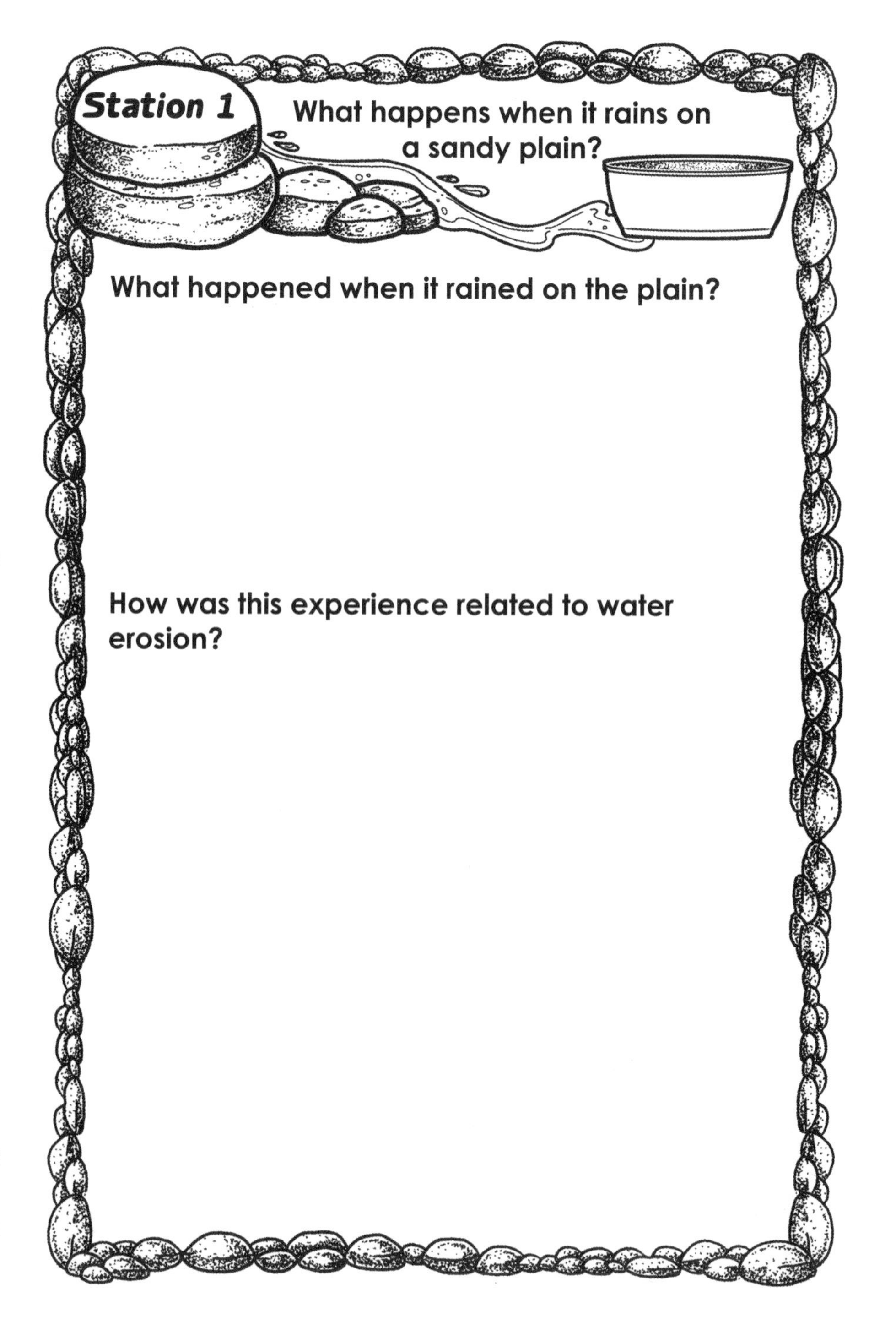

**Station 1**

**What happens when it rains on a sandy plain?**

**What happened when it rained on the plain?**

**How was this experience related to water erosion?**

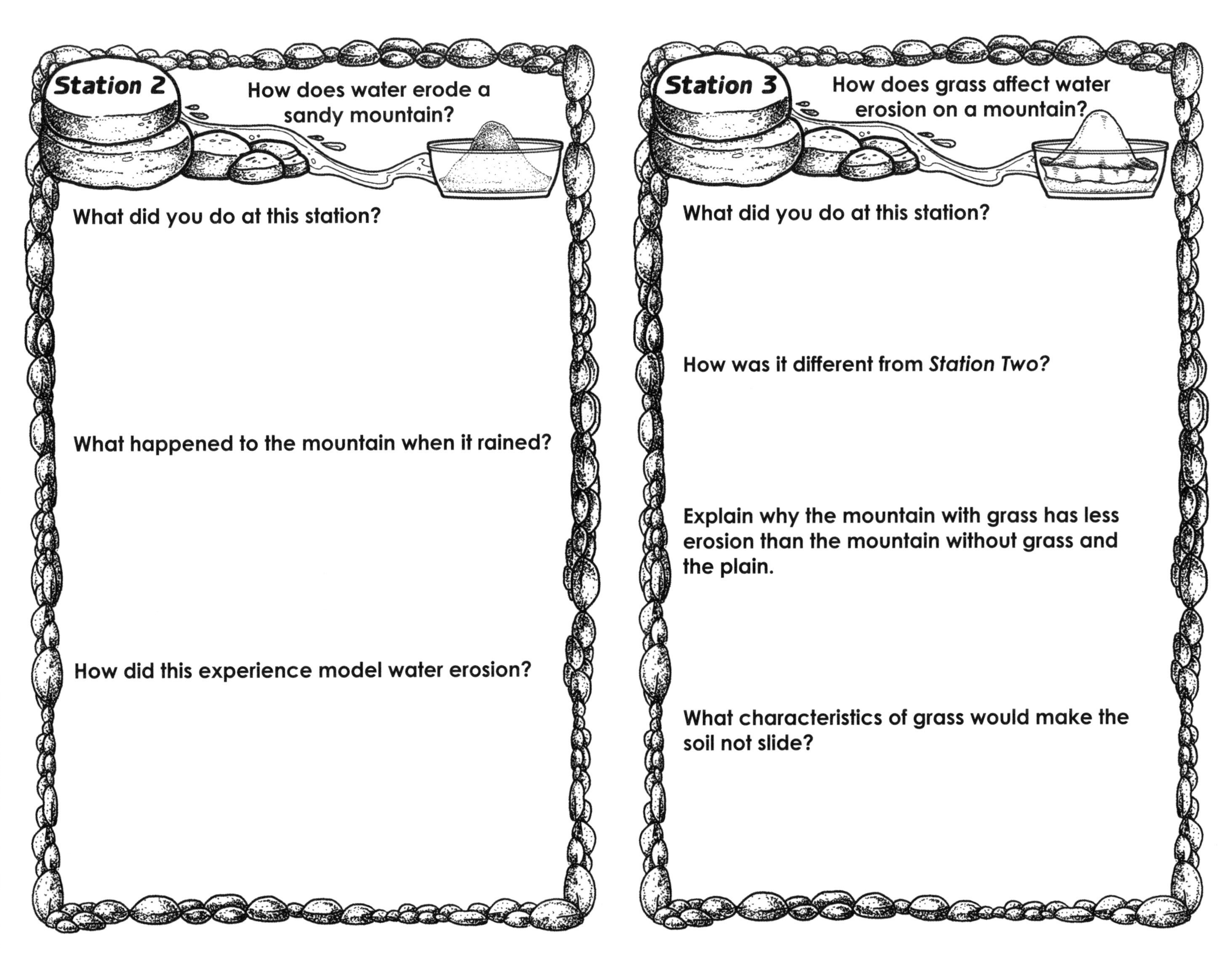

## Station 2

**How does water erode a sandy mountain?**

**What did you do at this station?**

**What happened to the mountain when it rained?**

**How did this experience model water erosion?**

## Station 3

**How does grass affect water erosion on a mountain?**

**What did you do at this station?**

**How was it different from *Station Two?***

**Explain why the mountain with grass has less erosion than the mountain without grass and the plain.**

**What characteristics of grass would make the soil not slide?**

 

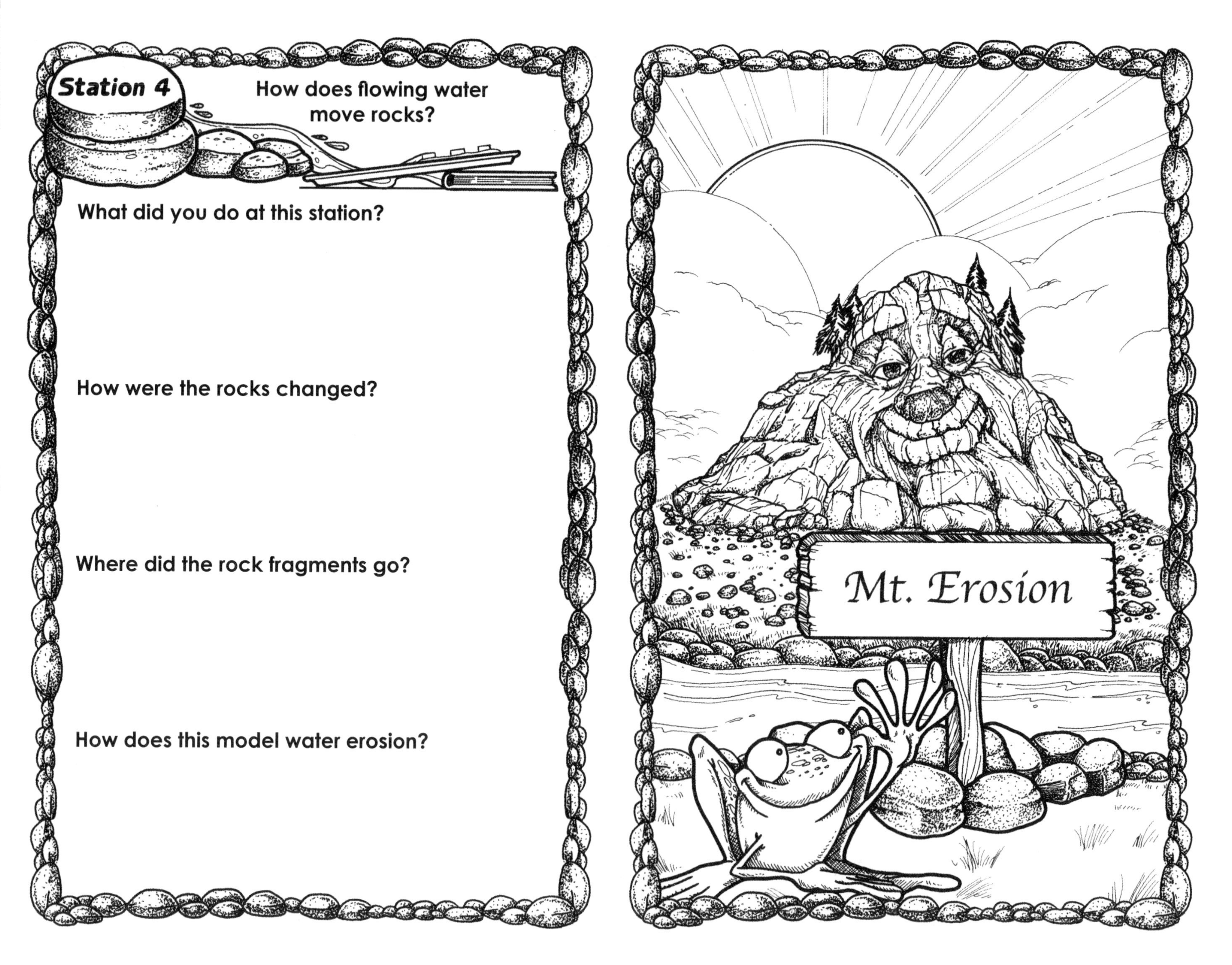
Station 4
How does flowing water move rocks?
What did you do at this station?
How were the rocks changed?
Where did the rock fragments go?
How does this model water erosion?
Mt. Erosion

## Connecting Learning

1. What is erosion?
2. What evidence of erosion did you see as you walked around the school?
3. How does water change the Earth?
4. How does wind change the Earth?
5. How does ice change the Earth?
6. What happened to the sand as you blew across your plot of land? How was the effect different when you added rocks and blew?
7. What are some examples of wind erosion in nature?

## Connecting Learning

8. How is the way a cave and a canyon are formed different? How is the process similar?

9. What role does water play in shaping the Earth?

10. Is a hurricane or tornado an agent of erosion? Explain your answer.

11. What are you wondering now?

 

# Peanut Butter and Jelly Geology

## Topic
Earth's crust

## Key Question
How do natural forces shape the rock layers of the Earth's crust?

## Learning Goals
Students will:
- make a model of rock layers in the Earth's crust, and
- use the model to learn how natural forces shape the rock layers.

## Guiding Documents
*Project 2061 Benchmarks*
- *Seeing how a model works after changes are made to it may suggest how the real thing would work if the same were done to it.*
- *Waves, wind, water, and ice shape and reshape the earth's land surface by eroding rock and soil in some areas and depositing them in other areas, sometimes in seasonal layers.*

*NRC Standards*
- *The surface of the earth changes. Some changes are due to slow processes, such as erosion and weathering, and some changes are due to rapid processes, such as landslides, volcanic eruptions, and earthquakes.*
- *Land forms are the result of a combination of constructive and destructive forces. Constructive forces include crustal deformation, volcanic eruption, and deposition of sediment, while destructive forces include weathering and erosion.*
- *Lithospheric plates on the scales of continents and oceans constantly move at rates of centimeters per year in response to movements in the mantle. Major geological events, such as earthquakes, volcanic eruptions, and mountain building, result from these plate motions.*

## Science
Earth science
  geology

## Integrated Processes
Observing
Comparing and contrasting
Applying

## Materials
*For each pair of students:*
- one slice of white bread
- one slice of whole wheat bread
- one slice of dark rye bread
- two tablespoons of jam or jelly
- two tablespoons of crunchy peanut butter mixed with raisins
- two paper plates
- jumbo craft stick or plastic knife
- plastic cafeteria gloves, optional

## Background Information
The surface of the Earth is undergoing change at all times. *Weathering* defines the disintegration and decomposition of the solid portions of the surface of the Earth. *Erosion* is the process of the movement of the Earth's materials that have been weathered.

Erosion can be divided into two components, *transport* and *deposition*. Transport is the movement of the weathered materials. The movement of these eroded materials is most often through water. Materials are often moved from one place and deposited in another location. The scientific term for this portion of the erosion process is deposition.

This activity models the processes of weathering, erosion, and deposition. Models are an important part of the study of Earth science. They allow us to learn about processes that are too slow or too large to observe.

## Management
1. Make sure that the sandwiches remain right side up while the students are making them (white bread on the bottom).
2. Have students work in pairs and share a sandwich.
3. It saves time to have the ingredients measured out onto paper plates before beginning this activity. A pair of students will share one paper plate of ingredients.
4. If students wear gloves, the sandwiches can be eaten afterward.
5. Check for allergies to peanuts. Other nut butters can be substituted.

**Procedure**

1. Distribute a paper plate with the described ingredients.
2. Tell the students you will show them how to make and manipulate a sandwich in the same way that natural forces shape layers of rock.
3. Use the accompanying narrative and pictures to provide a guide for building the sandwich.
4. As the students build their sandwiches, keep track of their progress by drawing a diagram on the board or referring to pictures on the activity sheet.
5. When all the sandwiches are finished, start a question and answer period. (See *Connecting Learning*.)

*Optional Investigations*

1. When geologists study layers of rock, they rarely find them flat and horizontal. Often they will see layers that are bent or broken. To illustrate these structures, have the students gently bend their sandwich to form a hill (always keeping the oldest layer on the bottom). This is called an *anticline.* Have the students bend the sandwich to form a trough. They now have a *syncline.* Mountains and valleys are formed in this way.
2. Sometimes the crust of the Earth moves up or down. In part, this movement causes earthquakes. Tell students to cut their sandwiches in half and move one half up or down. Tell them to hold up the two halves. Which side moved up or down? Either the left side moved up and the right side moved down or the other way around. Ask the students how they can tell from the model that the Earth moved. [The layers don't line up anymore.] Inform the students that this is a *vertical fault.* Have them draw this on their papers.

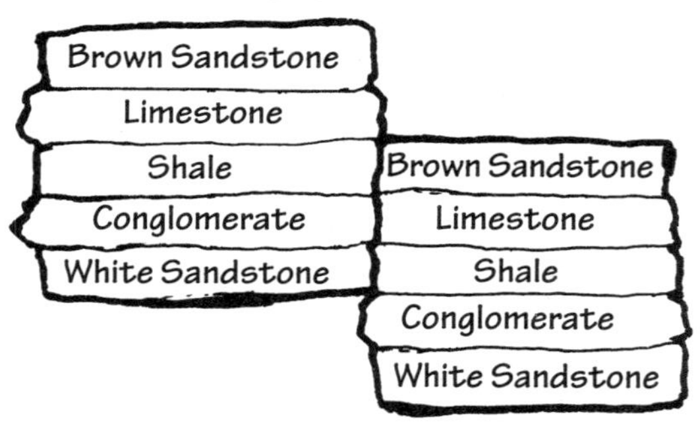


3. Tell the students that they are going to observe a lateral fault. Show them how to slide the two parts of the sandwich past each other on the same level.

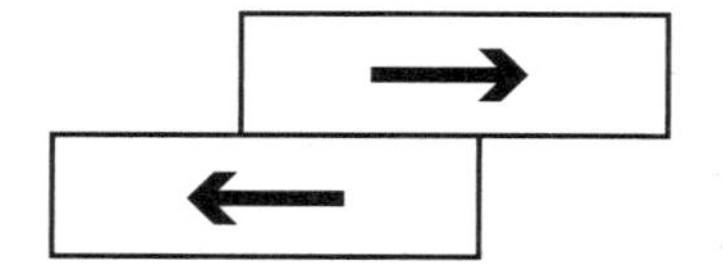

**Connecting Learning**

1. It took about 10 minutes for all pairs to build their sandwiches. What is the oldest part? [bottom layer] Why do you say that? [It was put on first; everything else went on top of it.]
2. What is the youngest layer of the sandwich? [the top layer] Why do you say that? [It was the last thing put on.]
3. What is the age of the shale or wheat bread? [Somewhere between the oldest and youngest.]
4. Why can't we say that it is half as old as the oldest layer and twice as old as the top layer? [We don't know how long it took for each layer to be added.]
5. What is the best way to determine the age of the limestone layer? ...the conglomerate layer? In any sandwich that is right side up (or a layer not overturned), how can we best describe the age of any particular layer? [Any one layer is younger than what is under it and older than what is on top of it.]
6. What processes built our rock layers? [erosion by water and wind]
7. What are you wondering now?

# Peanut Butter and Jelly Geology

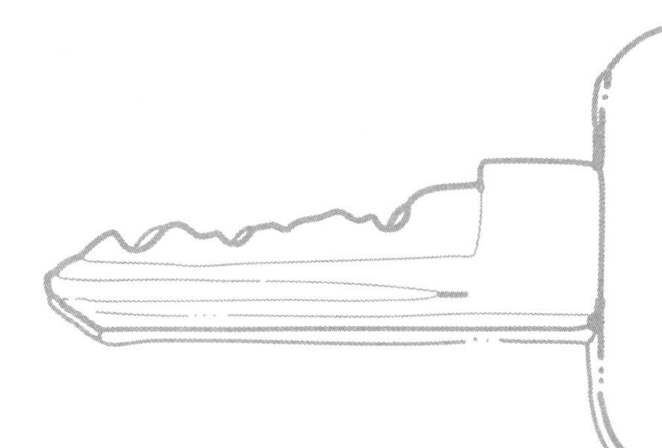

## Key Question

How do natural forces shape the rock layers of the Earth's crust?

## Learning Goals

**Students will:**

- make a model of rock layers in the Earth's crust, and
- use the model to learn how natural forces shape the rock layers.

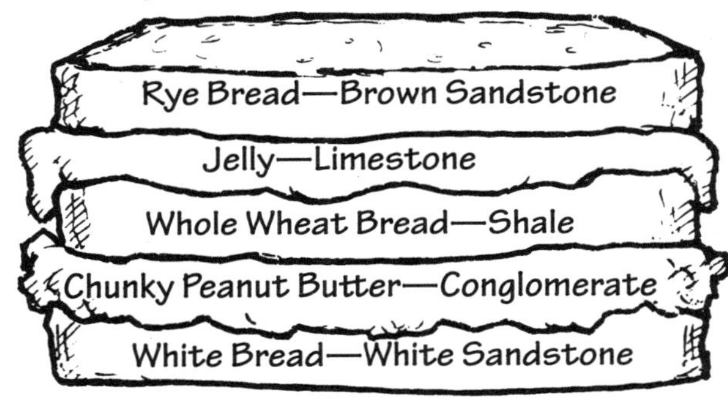


 

# Peanut Butter and Jelly Geology

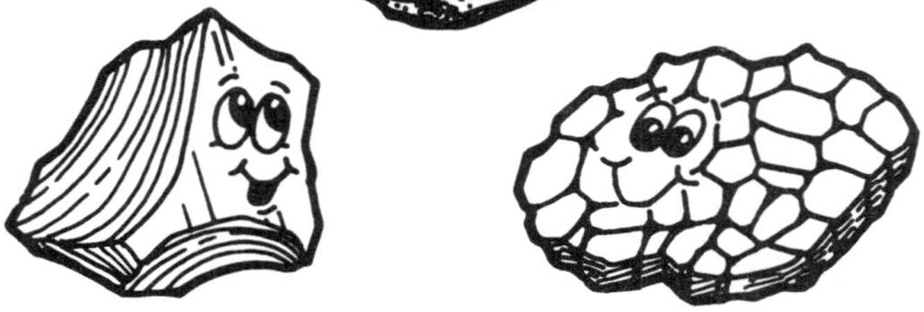

1. Here we have an area of igneous bedrock. A river is flowing swiftly along, carrying a load of white sand eroded from white rocks some distance away. The river channel gets wider, so the river slows down as it spreads out. The more slowly the water moves, the less load it can carry, so it drops its load of white sand along the bottom. As the years pass, this sand becomes cemented together, forming a layer of white sandstone.

(white bread)

2. A major flood occurs. Tons of mud, rocks of all sizes, and debris come pouring through this area and cover the white sand. The mud is represented by peanut butter; chunky, because of all the rocks it carries. The raisins are the boulders wrenched loose by the flood.

(chunky peanut butter and raisins)

3. In time, a slower, more gentle river flows across the area. Even though it's a calmer river, it still carries a load. It has traveled through clay and is carrying a large amount of fine clay particles known as silt. As the river slows down and becomes shallower, this silt is dropped to the bottom, forming another layer. In time, this silt will become the sedimentary rock called shale.

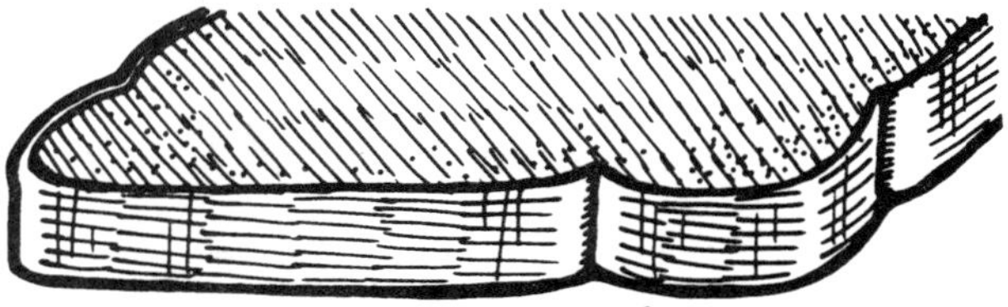

(wheat bread)

4. Meanwhile, the Earth is warming and glaciers are melting. The ocean rises and covers the area we are observing. The salt water brings with it millions of little marine organisms. Their shells begin to line the ocean floor, forming a new layer. In time, the ocean will recede, leaving a calcium-rich layer that will become limestone.

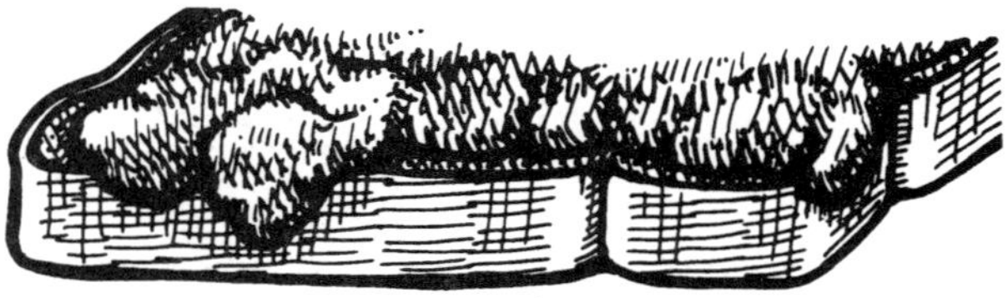

(jelly)

5. It is a time of severe drought. Mighty winds pick up eroded bits of brown rocks. These rocks sandblast other rocks, until finally a layer of brown sand covers the area we are observing. In time, this layer will become brown sandstone.

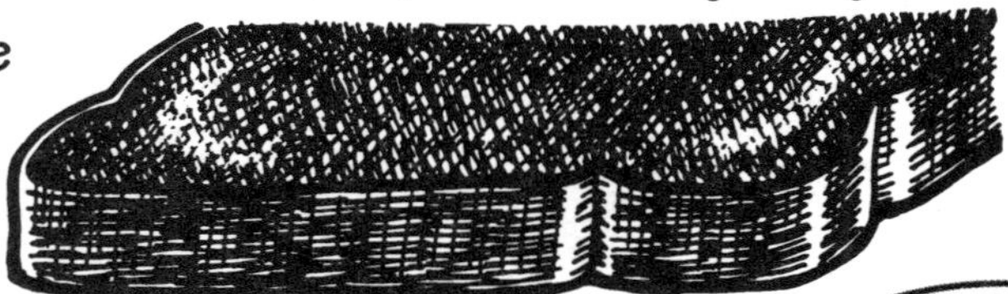

(dark bread)

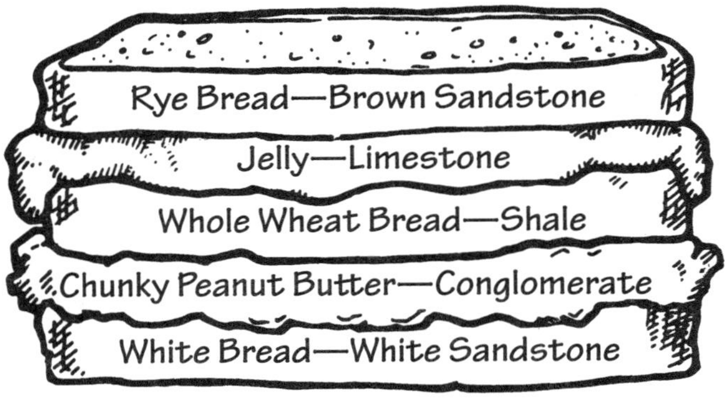
Rye Bread—Brown Sandstone
Jelly—Limestone
Whole Wheat Bread—Shale
Chunky Peanut Butter—Conglomerate
White Bread—White Sandstone

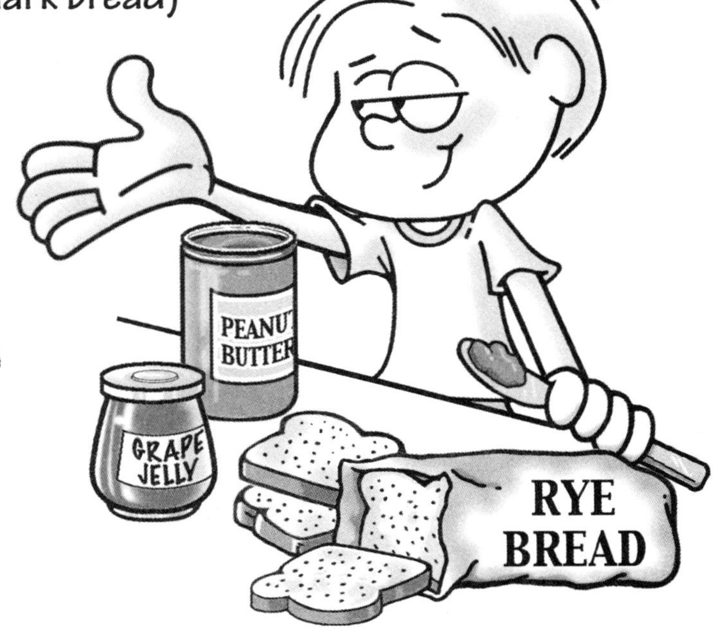
PEANUT BUTTER
GRAPE JELLY
RYE BREAD

# Peanut Butter and Jelly Geology

Follow the directions found in the story to answer this question.

**How do natural forces shape the rock layers of Earth's crust?**

Sandwich Layers

Corresponding Layers of Rock

Use this space to describe the processes of weathering, transport, and deposition as illustrated in this activity.

# Peanut Butter and Jelly Geology

CONNECTING LEARNING

## Connecting Learning

1. It took about 10 minutes for all pairs to build their sandwiches. What is the oldest part? Why do you say that?

2. What is the youngest layer of the sandwich? Why do you say that?

3. What is the age of the shale or wheat bread?

4. Why can't we say that it is half as old as the oldest layer and twice as old as the top layer?

# Peanut Butter and Jelly Geology

CONNECTING LEARNING

## Connecting Learning

5. What is the best way to determine the age of the limestone layer? ...the conglomerate layer? In any sandwich that is right side up (or a layer not overturned), how can we best describe the age of any particular layer?

6. What processes built our rock layers?

7. What are you wondering now?

 

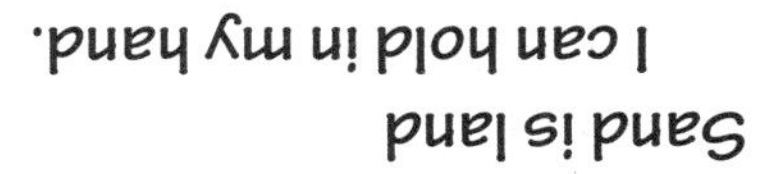

Sand is land
I can hold in my hand.

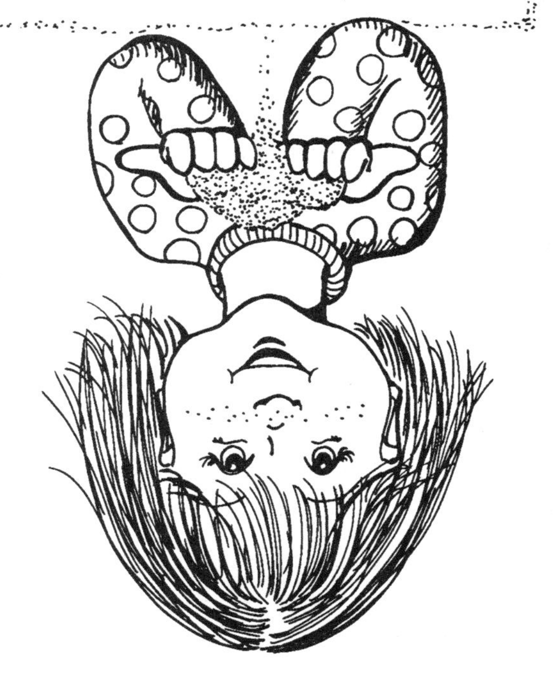



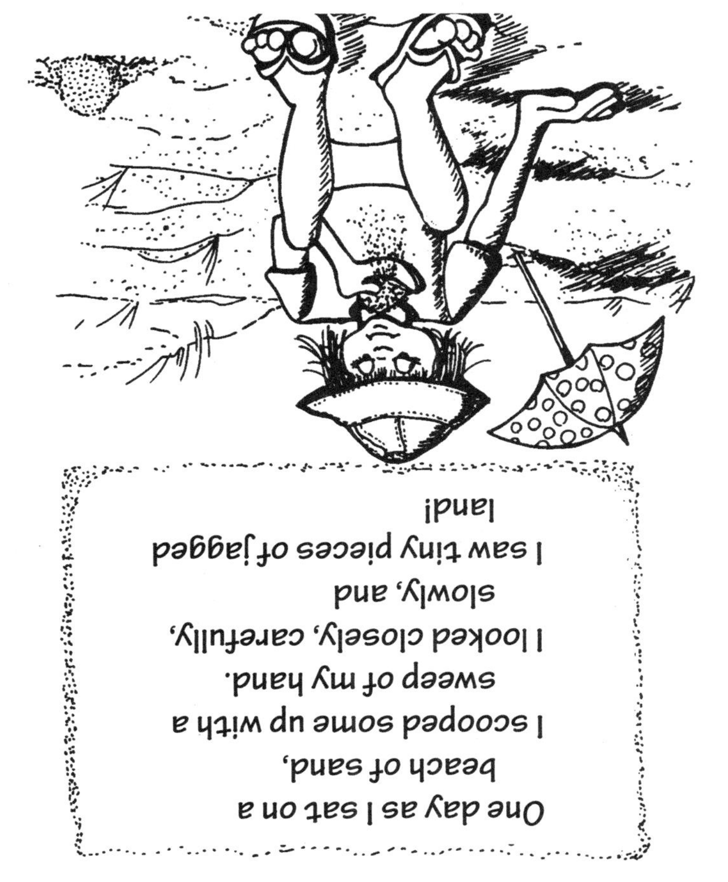

One day as I sat on a
beach of sand,
I scooped some up with a
sweep of my hand.
I looked closely, carefully,
slowly, and
I saw tiny pieces of jagged
land!

Reader's Review

This book made me think about

______________________

______________________

______________________

______________________

______________________

______________________

______________________

Reviewed by: ______________

12

This book belongs to:

1

 

9

4

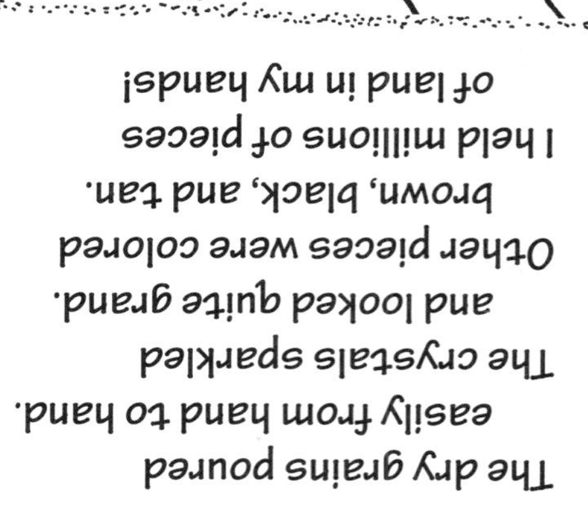


Billions and billions of
grains of sand
Stretched out before me
across the strand.
Each sparkling grain
came from land
Far, far away from the
beach of sand.

10

Sand is land
I can hold in my hand.

3

 

7

Sand is land
I can hold in my hand.

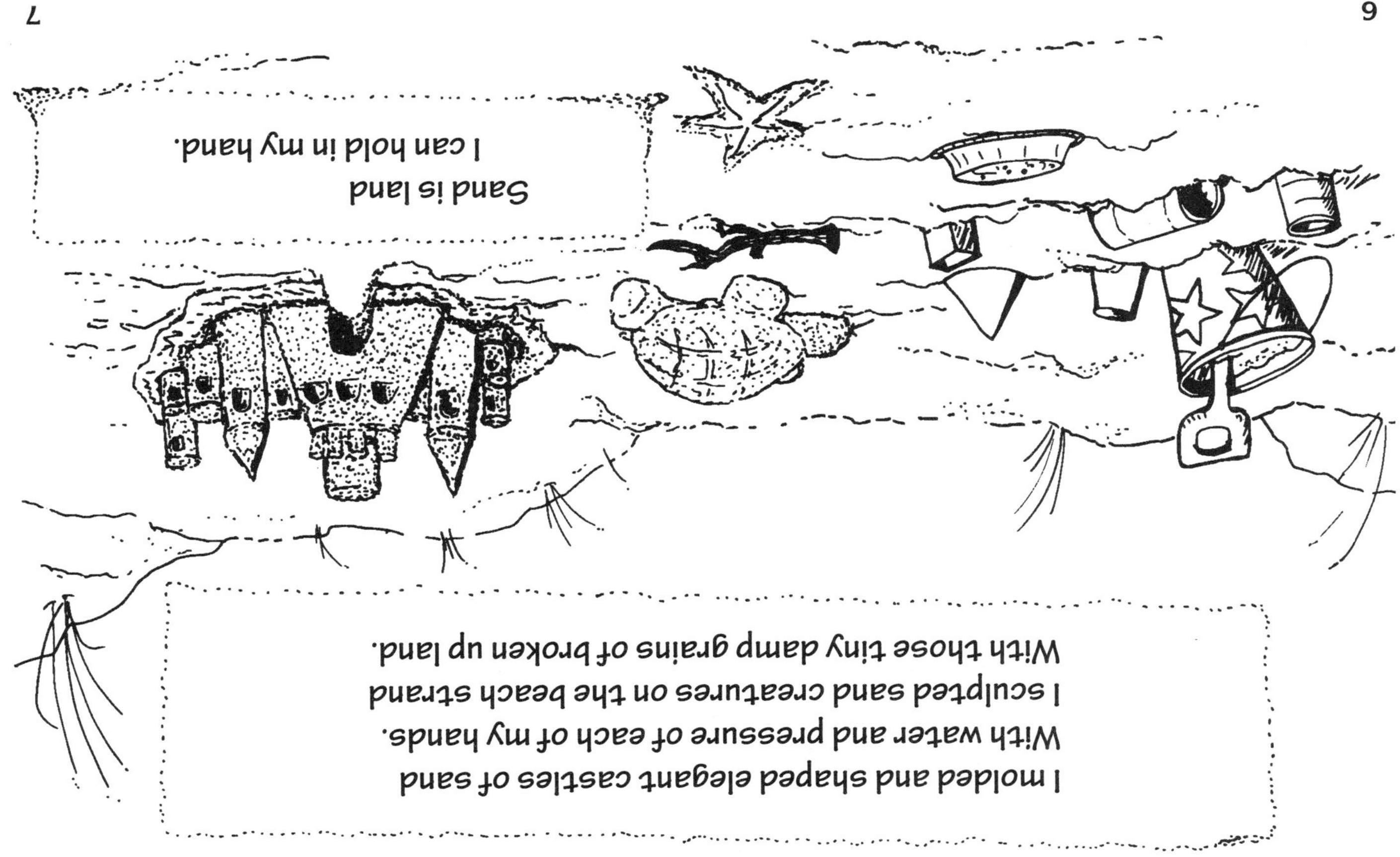

I molded and shaped elegant castles of sand
With water and pressure of each of my hands.
I sculpted sand creatures on the beach strand
With those tiny damp grains of broken up land.

The waves rolled in and
washed the strand
Of castles and creatures
I sculpted with sand.
I moved up the beach
where I could stand
To look over that changing
land of sand.

8

Sand is land
I can hold in my hand.

5

7

Soil and rocks are renewable resources. Forces of the Earth make rocks. Lava from volcanoes makes new rocks. The pressure of the Earth on sand also makes rocks.

2

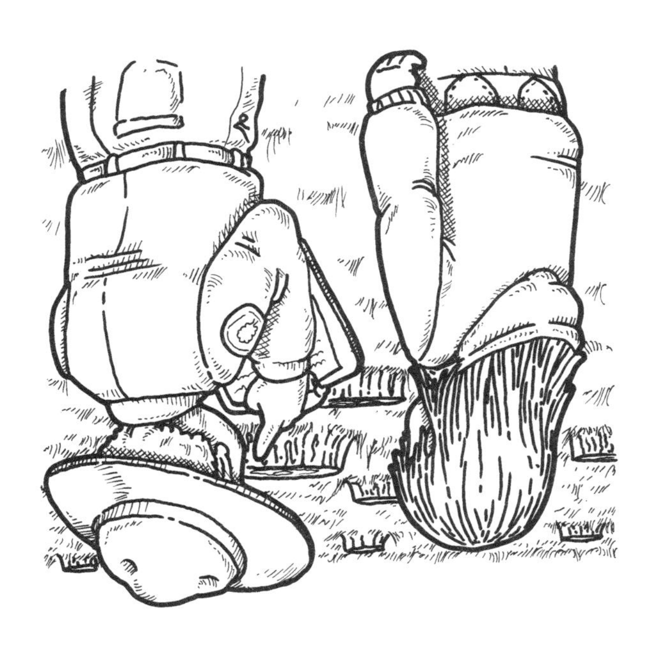

Resources that are renewable can be replaced.

Draw a picture of a resource that is renewable.

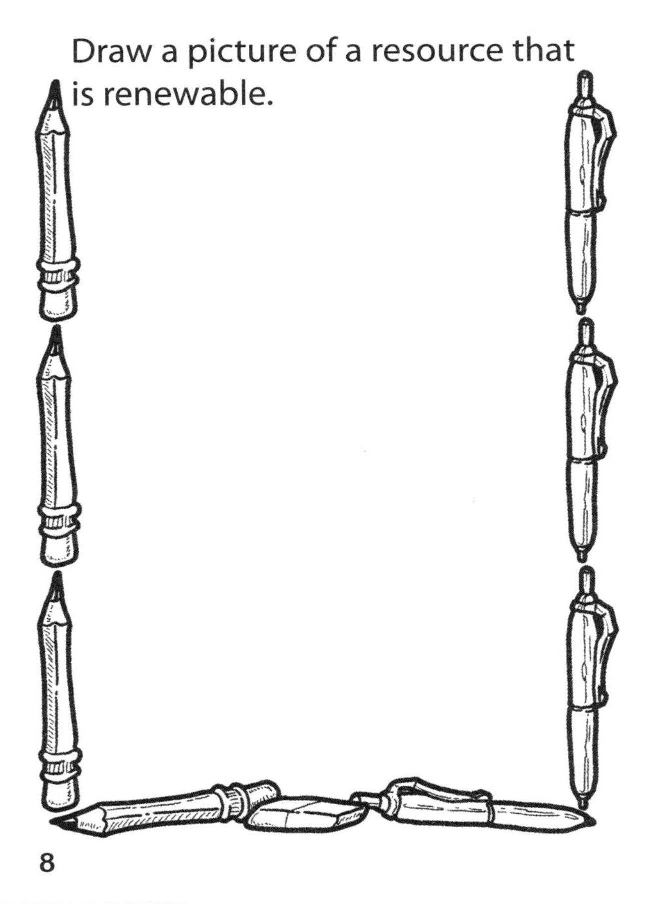

8

# Renewable RESOURCES

1

Plants are renewable resources. When a tree is cut down, another one can be planted to replace it.

3

Animals are considered renewable resources. They have babies.

4

Thanks to the water cycle, water is a renewable resource. It rains. The water evaporates and makes clouds. Then it rains again.

5

Gases in the air are also renewable. Carbon dioxide and oxygen are two gases found in the air. Plants use carbon dioxide and give us oxygen. People use oxygen and breathe out carbon dioxide.

6

 

# Nonrenewable RESOURCES

Nonrenewable resources cannot be replaced.

1

We use these resources faster than the Earth can make them. Minerals are nonrenewable resources. Gold, silver, and copper are minerals.

2

We use oil, natural gas, and coal faster than they can be replaced. They are nonrenewable resources.

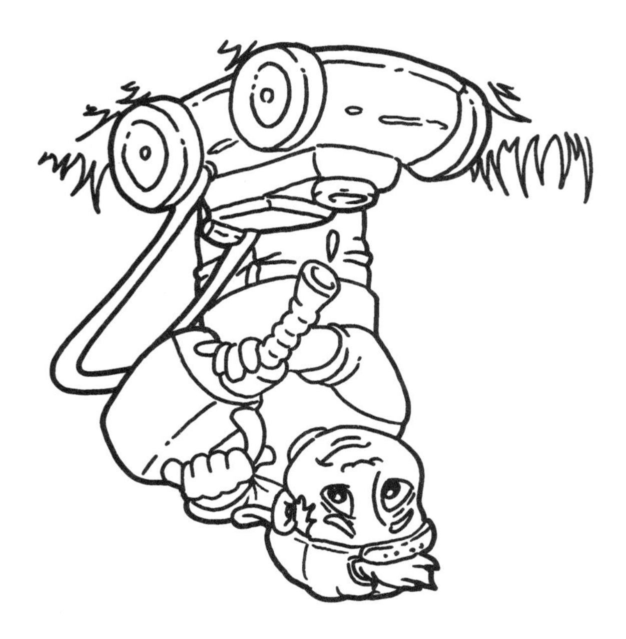

3

We can help save our nonrenewable resources. When we use our cars less, we are helping save our resources. When we turn off the lights that we don't need, we are helping save our resources.

4

**Topic**
Earth's resources

**Key Question**
Which of these resources are renewable and which are nonrenewable?

**Learning Goal**
Students will read clues to determine which resources are renewable, and which are nonrenewable.

**Guiding Document**
*NRC Standards*
- *Earth materials are solid rocks and soils, water, and gases of the atmosphere. The varied materials have different physical and chemical properties, which make them useful in different ways, for example, as building materials, as sources of fuel, or for growing the plants we use as food. Earth materials provide many of the resources that humans use.*
- *Resources are things we get from the living and nonliving environment to meet the needs and wants of a population.*
- *Some resources are basic materials, such as air, water and soil; some are produced from basic resources, such as food, fuel, and building materials; and some resources are nonmaterial, such as quiet places, beauty, security, and safety.*

**Science**
Earth science
    resources
        renewable, nonrenewable

**Integrated Processes**
Observing
Comparing and contrasting
Drawing conclusions

**Materials**
Glue sticks
Scissors
Picture cards
Rubber bands, #19
*Recognizing Resources* rubber band book

**Background Information**
Renewable resources are those things that can be replaced in a short amount of time. Even though resources like plants, animals, water, air, and soil can be renewed, they must be used wisely. Otherwise, they will not be available for our long-term use. Wise decision-making can take many forms from reducing pollution that damages our air and water to protecting our wildlife species from over-hunting or from the destruction of their habitats.

Nonrenewable resources are those things that cannot be replaced in a short amount of time. Fossil fuels and minerals are nonrenewable resources. It takes millions and millions of years to form oil, natural gas, and coal. We use them much faster than they can be replenished. Conservation is the key. Carpooling, walking, riding bikes, combining trips, turning off unused lights, adjusting the thermostats, etc., are easy ways in which we can conserve our nonrenewable resources.

**Management**
1. Each student should make his or her own rubber band book.
2. Time should be spent developing the vocabulary (renewable, nonrenewable) while doing this activity. When students understand the roots of the words, the application to resources is much easier.

**Procedure**
1. Ask the *Key Question* and state the *Learning Goal.*
2. Distribute the page of pictures, scissors, and a glue stick to each student.
3. Tell the class they will be reading descriptions of resources found on Earth. Once they determine what is being described, they will cut out the appropriate picture and glue it onto the page of their rubber band book.
4. Distribute the pages of the book and the rubber bands. Have students fold and nest the pages and secure them with the #19 rubber bands.
5. When students are finished, conclude with a discussion of the classification of resources.

**Connecting Learning**
1. Which of Earth's resources did you use today?
2. Were they renewable or nonrenewable? Explain.
3. Why is it important to protect our resources?
4. What are some things we can do to protect our resources?
5. What resources do you think people waste? How are they wasted?
6. What are you wondering now?

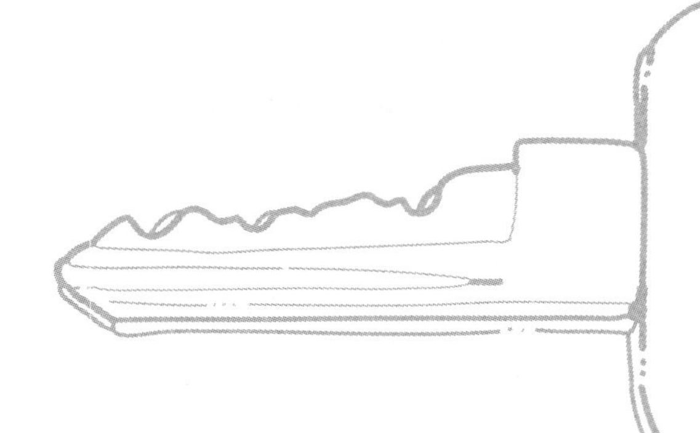

**Key Question**

Which of these resources are renewable and which are nonrenewable?

## Learning Goal

read clues to determine which resources are renewable, and which are nonrenewable.

Recognizing Resources

$000
$000
Coal

I am a **nonrenewable resource**. People use me as fuel for their cars. It is important to use me wisely.

I am a **renewable resource**. Plants and animals need me to survive. I contain oxygen, carbon dioxide, and other gases.

I am a **nonrenewable resource**. People use me to heat their homes and power their stoves.

**Renewable resources** are resources that can be replaced. We need to take care of them. **Nonrenewable resources** cannot be replaced in a short period of time. We must use them wisely.



 

5

I am a **renewable resource.** I have babies to make more like me. People eat me for food. They use my hide for clothing.

4

I am a **renewable resource.** I am found in lakes, rivers, and oceans. I am essential to life on Earth.

I am a **nonrenewable resource**. I am mined from the ground and burned in power plants. I am used to produce electricity for your homes.

6

I am a **renewable resource**. I am used to build houses. When I am cut down, people plant more just like me. The new ones will be used to make more houses.

3

CONNECTING LEARNING

## Connecting Learning

1. Which of Earth's resources did you use today?
2. Were they renewable or nonrenewable? Explain.
3. Why is it important to protect our resources?
4. What are some things we can do to protect our resources?
5. What resources do you think people waste? How are they wasted?
6. What are you wondering now?

# Classifying Resources

**Topic**
Earth's resources

**Key Question**
Which of Earth's resources are renewable, and which are nonrenewable?

**Purpose of the Game**
Students will match pictures of Earth's resources to the proper category.

**Materials**
Game cards

**Management**
1. This game does not introduce the classification of Earth's resources, so be sure that students have learned the categories and examples before doing this activity.
2. Each group of two to four students will need one set of playing cards. The cards should be copied onto card stock and laminated, if possible.

**Procedure**
1. Have students get into their groups and distribute one set of playing cards to each group.
2. Instruct students to shuffle the cards and lay them face down in a grid.
3. Explain that one person will begin by turning over any two cards. If those cards both show a picture of a resource and the proper category for it, the player gets to remove the cards and turn over two more. If the cards do not match, it is the next player's turn.
4. Tell them to continue playing until all of the cards have been paired.
5. When groups have completed their games, the student with the most pairs wins.
6. Play as desired to reinforce the classification of Earth's resources.

# Classifying Resources

| | | |
|---|---|---|
| Renewable Resource | Renewable Resource | Renewable Resource |
| Renewable Resource | Renewable Resource | Nonrenewable Resource |
| Nonrenewable Resource | Nonrenewable Resource | Nonrenewable Resource |

 

# Classifying Resources

## Topic
Natural resources

## Key Question
From what natural resource do these items come?

## Learning Goals
Students will:
- research the natural resources from which a variety of items come, and
- identify these resources as renewable or nonrenewable.

## Guiding Document
*NRC Standards*
- *Earth materials are solid rocks and soils, water, and gases of the atmosphere. The varied materials have different physical and chemical properties, which make them useful in different ways, for example, as building materials, as sources of fuel, or for growing the plants we use as food. Earth materials provide many of the resources that humans use.*
- *Resources are things we get from the living and nonliving environment to meet the needs and wants of a population.*
- *Some resources are basic materials, such as air, water and soil; some are produced from basic resources, such as food, fuel, and building materials; and some resources are nonmaterial, such as quiet places, beauty, security, and safety.*

## Science
Earth science
  natural resources

## Integrated Processes
Observing
Collecting and recording data
Classifying
Comparing and contrasting

## Materials
*For each group:*
  project description page
  recording page, one per student
  colored pencils, red and green

*For the class:*
  *Agatha's Feather Bed* (see *Curriculum Correlation)*
  picture cards (see *Management 2)*
  research sources (see *Management 5)*

## Background Information
Natural resources are things, such as plants and animals, rocks and minerals, or fresh water, that are found in nature and are necessary or useful to humans. These resources can be renewable or nonrenewable. These resources are essential to our survival. Without fresh water to drink and food from plants and animals, we could not survive. Our homes are made from wood that comes from trees, our clothes are made from cotton that comes from plants, and our cars are made from steel that is made from rocks and minerals. Everything we use and make comes from the Earth.

In this activity, students will research the materials that compose several common items and identify the natural resources from which these materials come. For example, a crayon is made of paraffin, a kind of wax. Wax is a petroleum product that is made from oil. The paper wrapper on a crayon is made from wood pulp that comes from trees. Once students have identified the resources from which an item is made, they will classify those resources as renewable or nonrenewable. The emphasis is not on thoroughly examining the origins of each material but on recognizing that humans need and use the Earth's natural resources, and that some of these cannot be renewed.

## Management
1. Students will need to work together in pairs (or groups of three, if necessary) to research the items provided. This research will need to take place over the course of several days to a week.
2. Copy the page of pictures and cut apart the cards. Put the pictures in a hat or bag from which groups can select the item they will research.
3. Plan to display the page of pictures using a projection device when introducing the activity.
4. The scope of the report is not defined on the student page other than by giving students questions to guide their research. You will need to decide on the format students will use and the expectations you have for them. For example, you may choose to have students complete the reports in a poster format and require that they include pictures of the natural resources. Be sure to clearly outline these expectations before students begin.
5. It is strongly recommended that you obtain at least one of the reference books listed in *Curriculum Correlation* for students to use as part of their research. Other research can be done on the Internet. See *Internet Connections* for two suggested websites.

 

6. The intent of this activity is not for students to do an in-depth analysis of the components of the items they select. Every effort has been made to select items with simple components that are easily identifiable. For example, pencils are typically wood, graphite, rubber, and metal. The fact that some rubber erasers are synthetic rather than natural is not important. Students should still recognize that natural rubber comes from trees and should identify the pencil as being made from plants and animals (as well as rocks and minerals).

**Procedure**

*Part One*

1. Read the book *Agatha's Feather Bed* to the class. Discuss the idea that nothing comes from nothing, and ask students to identify some of the natural resources that were mentioned in the book.
2. Write the categories of natural resources on the board: *Plants and animals; Rocks and minerals; Fresh water;* and *Oil, coal, and natural gas.* Discuss some of the specific things that would come under each of these categories. For example, trees would fall under *Plants and animals,* and metals like iron and aluminum would be considered *Rocks and minerals.*
3. Display the page of pictures. Explain that the pictures show things they would come across at home or at school that are made from natural resources. It will be their job to do some research on one of these things.
4. Have students get into pairs and distribute the page describing the research project. Move from group to group and allow each pair to select one of the pictures from the hat or bag.
5. Go over the instructions as a class. Discuss the information that students will need to collect and explain your expectations for how that will be compiled in a report.
6. Inform students of the amount of time they will have to complete the project and set the time for them to share their findings with the rest of the class.

*Part Two*

1. Distribute the student page for *Part Two* to each student. Have them record the names of their items under each heading that is appropriate. For example, *pencil* should be recorded both under *Plants and animals* (for the wood and rubber) and *Rocks and minerals* (for the metal and graphite).
2. Allow time for groups to share their projects with the class. As groups share, have students complete the student page by recording all of the items under every resource that is appropriate.
3. When groups are done, discuss the natural resources that are used most in these examples and whether they are renewable or nonrenewable. Have students use the colored pencils to underline the resources that are renewable in green and those that are nonrenewable in red.

**Connecting Learning**

*Part One*

1. What are some of the natural resources in the story we read? [silk, cotton, goose down, etc.]
2. Which of these things would be natural resources from plants and animals? ...rocks and minerals? ...etc.?
3. Which of the natural resources listed do you think we use the most? Justify your response.
4. Which kind of natural resource is used to make paper? [plants and animals] ...paper clips? [rocks and minerals] How do you know?

*Part Two*

1. Of what materials is your item made?
2. What natural resources are the sources of these materials?
3. Are the resources used to make your item renewable, nonrenewable, or both?
4. What interesting things did you learn about your item and the natural resources from which it is made?
5. Which category of natural resources was used the most in making the items we researched? ...the least?
6. Were these resources renewable or nonrenewable?
7. Do you think the same categories of natural resources would be most and least common if we looked at other items? Explain.
8. Does knowing the resources used to make any of the items change how you think about using them? Explain. [For example, plastics are made from petroleum—a nonrenewable resource. Reducing the use of things like bottled water can help conserve this resource.]
9. What are you wondering now?

**Internet Connections**

*How Products Are Made*
http://www.madehow.com/
This site explains the manufacturing process of a wide variety of items from blue jeans to nuclear submarines.

*How Everyday Things Are Made*
http://manufacturing.stanford.edu
This site, from the Product Realization Network at Stanford University, has videos that explain the manufacturing process for products in categories such as transportation, what you wear, and packing it up.

 

**Curriculum Correlation**

Deedy, Carmen Agra. *Agatha's Feather Bed: Not Just Another Wild Goose Story*. Peachtree Publishers. Atlanta, GA. 1991.

Rose, Sharon and Neil Schlager. *How Things Are Made: From Automobiles to Zippers*. Black Dog & Leventhal. New York. 2003.

Slavin, Bill. *Transformed: How Everyday Things Are Made*. Kids Can Press. Toronto, Canada. 2005.

Smith, Penny and Lorrie Mack. *See How It's Made*. DK Publishing. New York. 2007.

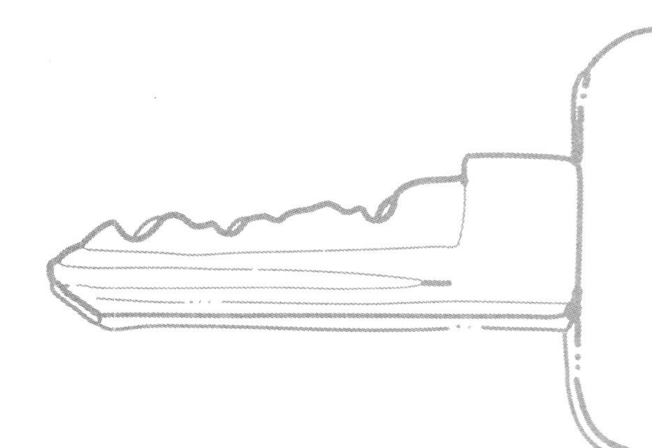

## Key Question

From what natural resource do these items come?

## Learning Goals

**Students will:**

- research the natural resources from which a variety of items come, and
- identify these resources as renewable or nonrenewable.

 

# Researching Resources

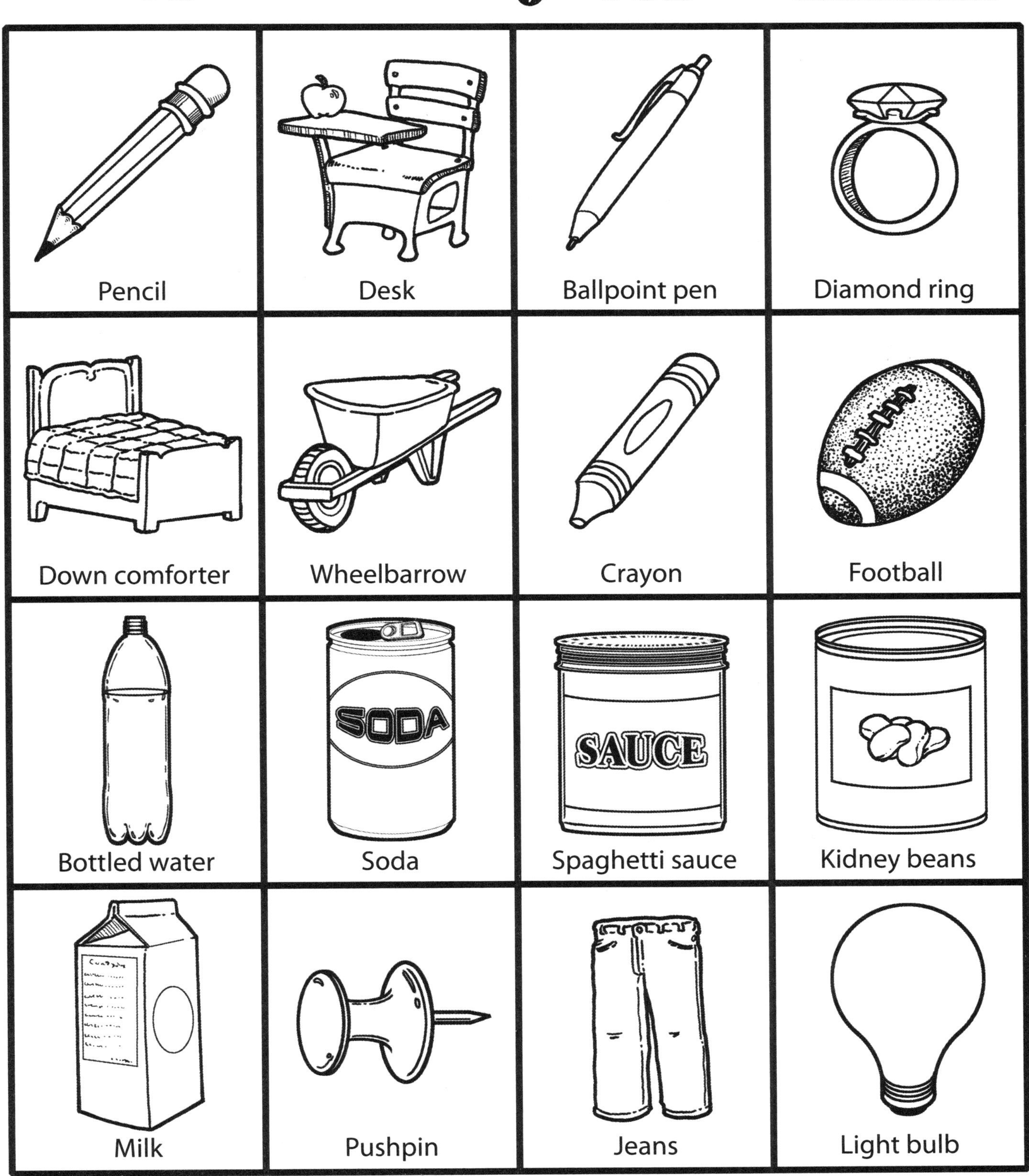

Your group is going to write a report about the item you selected. Your teacher will give you specific directions. Include the answers to these questions in your report:

- What are the materials from which your item is made? (For example, a car is made of steel, glass, rubber, plastic, etc.)
- What natural resource is the source of each of these materials? (For example, steel and glass come from rocks and minerals. Rubber comes from plants and animals. Etc.)
- Are the resources used to make your item renewable or nonrenewable?
- What is an interesting fact about your item and/or the natural resources from which it is made?

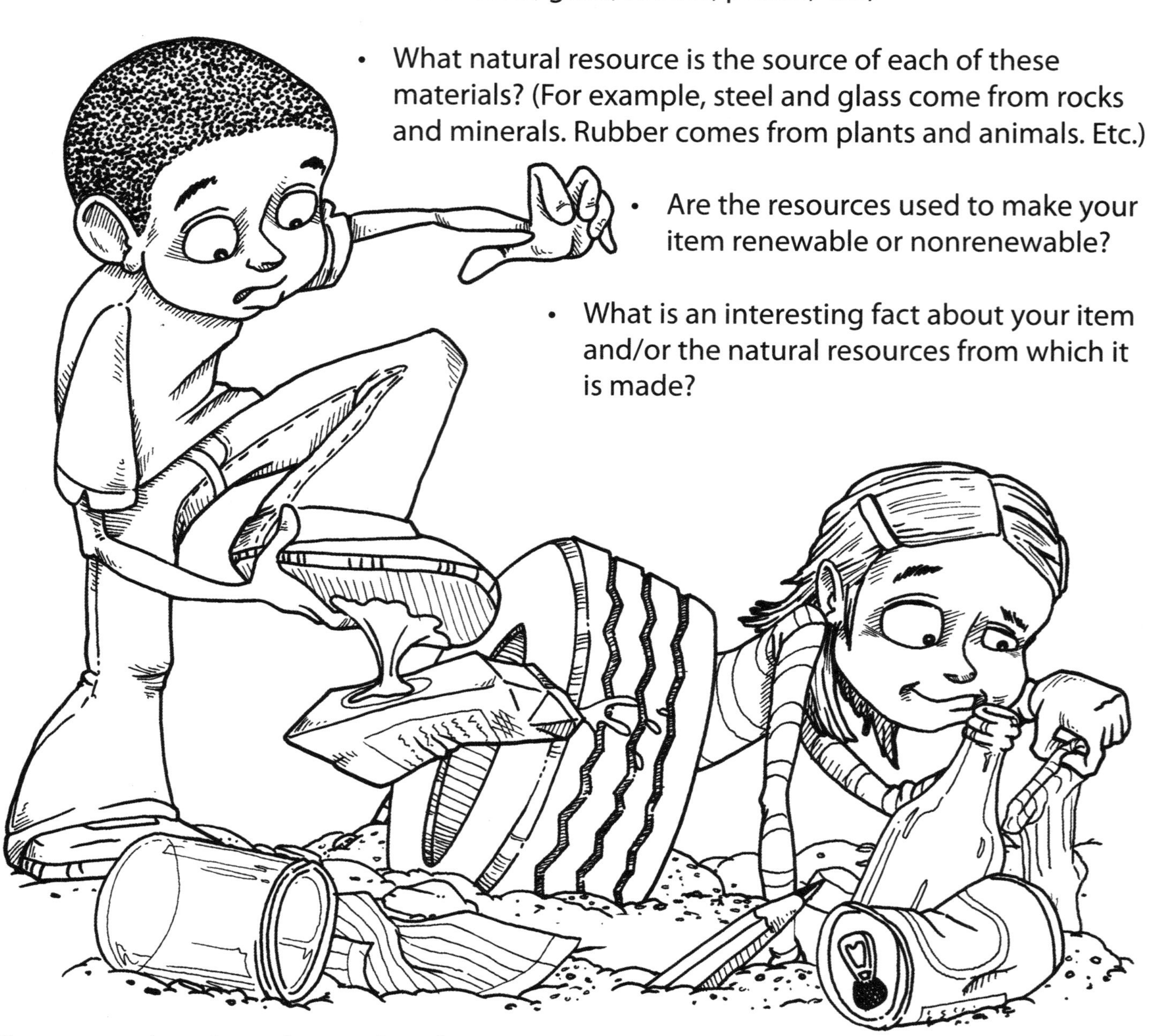

Be prepared to share the results of your research with the rest of the class.

Write the name of the item you researched under each natural resource from which it is made. Add the items your classmates researched to the lists as they share. Underline the resources that are renewable in green. Underline the resources that are nonrenewable in red.

## Connecting Learning

*Part One*

1. What are some of the natural resources in the story we read?

2. Which of these things would be natural resources from plants and animals? ...rocks and minerals? ...etc.?

3. Which of the natural resources listed do you think we use the most? Justify your response.

4. Which kind of natural resource is used to make paper? ...paper clips? How do you know?

## Connecting Learning

*Part Two*

1. Of what materials is your item made?

2. What natural resources are the sources of these materials?

3. Are the resources used to make your item renewable, nonrenewable, or both?

4. What interesting things did you learn about your item and the natural resources from which it is made?

CONNECTING LEARNING

## Connecting Learning

5. Which category of natural resources was used the most in making the items we researched? ...the least?

6. Were these resources renewable or nonrenewable?

7. Do you think the same categories of natural resources would be most and least common if we looked at other items? Explain.

8. Does knowing the resources used to make any of the items change how you think about using them? Explain.

9. What are you wondering now?

# Resource Relatives

**Topic**
Natural resources

**Key Question**
How do we use our natural resources?

**Learning Goal**
Students will identify ways that our natural resources can be used.

**Guiding Documents**
*Project 2061 Benchmarks*
- *Objects can be described in terms of the materials they are made of (clay, cloth, paper, etc.) and their physical properties (color, size, shape, weight, texture, flexibility, etc.).*
- *People can often learn about things around them by just observing those things carefully, sometimes they can learn more by doing something to the things and noting what happens.*

*NRC Standards*
- *Earth materials are solid rocks and soils, water, and gases of the atmosphere. The varied materials have different physical and chemical properties, which make them useful in different ways, for example, as building materials, as sources of fuel, or for growing the plants we use as food. Earth materials provide many of the resources that human use.*
- *Resources are things we get from living and nonliving environment to meet the needs and wants of a population.*
- *Some resources are basic materials, such as air, water, and soil; some are produced from basic resources, such as food, fuel, and building materials; and some resources are nonmaterial, such as quiet places, beauty, security, and safety.*

**Science**
Earth science
natural resources

**Integrated Processes**
Observing
Classifying
Relating

**Materials**
*For the class:*
*The Little Red Hen* by Paul Galdone

*For each group:*
paper lunch bag
natural resource cards
scratch paper
items made from natural resources (see *Management 4)*
game cards
game pieces (see *Management 6)*

**Background Information**
The Earth provides everything people need. Our food, materials needed to make clothing and shelter, and even the fuels that furnish us energy come from the Earth. These things that we get from the Earth are considered natural resources. They can be material resources (e.g., plants, animals, soil, minerals, rocks, water) or energy resources (e.g., solar, fossil fuels, heat, light, wind, geothermal, etc.). In this activity, students will become familiar with what natural resources are used to make specific products.

**Management**
1. Student should work together in groups of four. If necessary, some groups can have five students.
2. Make copies of the natural resource cards on card stock and laminate for extended use. Cut them apart and place them in a paper lunch bag. Each group needs one bag of cards.
3. Make sure each group has scratch paper to keep score of points earned during the game in *Part One.*
4. For *Part Two* of the activity, gather the following items:
   - Animals: wool scarf/sweater, leather glove/belt, pearl-like necklace/earring, silk scarf, honey
   - Plants: cotton towel, cardboard box, linen napkin/shirt, perfume, wooden ruler, paper
   - Rocks/Minerals: salt, glass baby food jar, brick, ceramic tile, clay pot
   - Metals: aluminum foil or can, nail, nickel, pan
   - Fossil Fuels: plastic bag, plastic milk container or water bottle, nylons, petroleum jelly (Vaseline®)
5. There are two game cards given so that not all students will have the same card. Cut the page in half before distributing cards to students.

 

6. All students need several game pieces to cover their bingo cards. Beans, Unifix cubes, plastic tiles, etc., all work well as game pieces.

**Procedure**

*Part One*

1. Read the story of *The Little Red Hen.* Discuss the many steps that it took for the Little Red Hen to turn the natural resource—wheat grains—into something we need—bread.
2. Review or introduce what a natural resource is. [water, plants, animals, minerals, rocks, soil, etc.]
3. Tell the class that they are going to play a game called "How can we use it?" Explain that each group of four students will need to divide themselves into two teams.
4. Give each group a paper lunch bag with a set of natural resource cards. Demonstrate how one player from the first team will pick a natural resource card out of the bag and try to think of a way that the resource can be used. Explain that if he/she can give a legitimate way that the resource can be used, his/her team will get a point. Students will return the card to the bag after each turn, then the play moves to the opposing team. Remind students that if they pull a card that has already been pulled, they can't give the same use as mentioned before, they must think of another use for the item. Play continues until the allotted time runs out.
5. After playing the game for several minutes, gather the students and review some of the ways that the resources were used and certain characteristics that the resources might have that made them good for that particular use. For example, wood—used for houses because it is strong.

*Part Two*

1. Review that natural resources are gifts from nature that are used to make the things we need and want.
2. One at a time, display the items suggested in *Management 6.* Question students about what natural resource it came from. For example, the glass baby food jar can be traced back to sand. (Some items may require research or more discussion if students are not sure of their origin, such as plastic, cans, etc.) Discuss why certain resources are better suited for certain products. Wood is good for building houses and furniture because it is sturdy; glass, which comes from sand, is better for making windows than wood because glass is transparent.
3. Give each student a game card and game pieces and explain that they will be playing a game to see if they remember or can figure out what natural resource some of our everyday items come from.
4. Tell students that you will hold up an item and that they are to cover one space on their card that matches its origin. For example, if you hold up a silk scarf, they should cover the word *animal* because silk comes from silkworms.
5. End with a time for students to share what they have learned about where our products come from.

**Connecting Learning**

1. How does the story of the Little Red Hen relate to our natural resources?
2. Why are some resources better than others when making certain products?
3. What are some ways that we use our natural resources? [building houses, making clothes, making machines, food]
4. What natural resources are necessary to produce all the materials that we used for our game? [sun, rocks, soil, grass, plants, etc]
5. What are you wondering now?

**Curriculum Correlation**

Galdone, Paul. *The Little Red Hen.* HMH Books. New York. 2011.

The Little Red Hen finds some grains of wheat on the ground, and asks for help in planting them. But her shiftless roommates—the dog, the cat, and the mouse—all refuse to help plant the wheat, water it, reap it, grind it, or bake a cake from the wheat. When the cake is ready to be eaten, they all want to help, but the hen eats the cake by herself. In the end, the lazy trio has learned to help with the household chores.

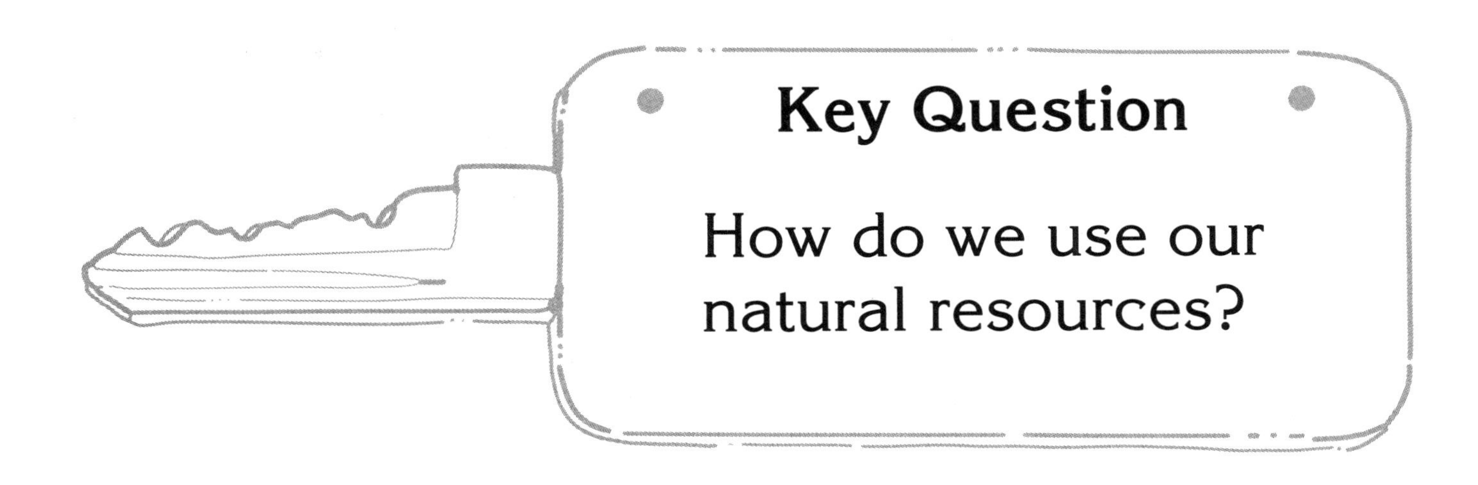


## Learning Goal

identify ways that our natural resources can be used.

 

# Resource Relatives

# Resource Relatives

## Game Card A

| | | | |
|---|---|---|---|
| Fossil Fuel | Animal | Metal | Plant |
| Rock/ Mineral | Plant | Animal | Fossil Fuel |
| Plant | Rock/ Mineral | Metal | Animal |
| Metal | Fossil Fuel | Plant | Rock/ Mineral |

# Resource Relatives

## Game Card B

| | | | |
|---|---|---|---|
| Animal | Rock/ Mineral | Plant | Metal |
| Plant | Animal | Fossil Fuel | Rock/ Mineral |
| Metal | Fossil Fuel | Plant | Animal |
| Rock/ Mineral | Metal | Fossil Fuel | Plant |

## Connecting Learning

1. How does the story of the Little Red Hen relate to our natural resources?

2. Why are some resources better than others when making certain products?

3. What are some ways that we use our natural resources?

4. What natural resources are necessary to produce all the materials that we used for our bingo game?

5. What are you wondering now?

# POLLUTION SOLUTION

**Topic**
Air pollution

**Key Question**
What was in the air that you breathed on your way to school today?

**Learning Goal**
Students will observe and identify pollutants in the air.

**Guiding Documents**
*Project 2061 Benchmark*
- *Human activities, such as reducing the amount of forest cover, increasing the amount and variety of chemicals released into the atmosphere, and intensive farming, have changed the earth's land, oceans, and atmosphere. Some of these changes have decreased the capacity of the environment to support some life forms.*

*NRC Standards*
- *When an area becomes overpopulated, the environment will become degraded due to the increased use of resources.*
- *Causes of environmental degradation and resource depletion vary from region to region and from country to country.*

**Math**
Estimation
Graphing

**Science**
Environmental science
airborne pollution

**Integrated Processes**
Observing
Comparing and contrasting
Collecting and recording data
Interpreting data
Classifying

**Materials**
Clear adhesive paper (see *Management 3)*
Tape
String
Non-corrugated cardboard (see *Management 4)*
Wax paper or plastic wrap
Petroleum jelly
Hand lenses

**Background Information**
Our normal daily activities release substances into the air. Some of those substances can cause problems for all occupants of this planet.

One type of pollution is a result of burning fuel for energy; this type is often called "black carbon" pollution. Small particles are released into the air. Diesel smoke is a visible form of this pollution. Automobiles, homes, and industries that burn fossil fuels are other major sources of black carbon pollution. Some environmentalists consider the burning of wood and charcoal to be a major source.

Noxious gases, such as sulfur dioxide, carbon monoxide, nitrogen oxides, and chemical vapors, comprise another type of pollution. These not only are pollutants in their own right, but often combine with other things in the atmosphere to form smog and acid rain.

Location has a lot to do with the amount of pollution in an area at a given time. Considerations as to geography, wind and weather factors, temperature, season of the year, vegetation, and population have to be made.

 

Outdoor pollution is one thing, but we also need to be concerned about indoor pollution. Eighty to ninety percent of our time is spent indoors. There are many sources of pollution that we encounter there: tobacco smoke; vapors from paint, carpets, and upholstered furniture; cooking and heating appliances; radon; etc.

There are short-term and long-term effects of pollution. They range from irritated eyes and throat to asthma, lung cancer, heart disease, etc.

**Management**

1. This activity will take approximately a week to do. A class period of 20-30 minutes is needed to choose working groups, make the collectors, and hang them where they will not be disrupted for three days. After three days, a 30-minute class period will be needed to retrieve and examine the collectors.
2. During the week, students could be assigned research projects in order to learn more about air pollution.
3. Clear adhesive paper, such as Contact™ paper can be purchased in many department stores, grocery stores, or home supply stores. Each pollution collector will need a 10-cm square.
4. Non-corrugated cardboard from shoeboxes or tablet backings work well. Each pollution collector will be mounted on an 11-cm square of cardboard.
5. Have some places in mind with walking distance of the school where the air-dirt collectors can be placed. It is suggested that one be placed near streets with a lot of traffic, another in a wooded or pastoral setting, and some indoors at home and school.
6. Have students work together in groups of four.
7. Prepare sample pollution collectors to serve as models for students. (See *Procedure 4.)*
8. Hand lenses (item number 1977) are available from AIMS.

**Procedure**

*Day One*

1. Ask the *Key Question* and state the *Learning Goal.*
2. Hold a discussion about what students know about air pollution. Have them discuss and record what things they think add to the pollution of our air.
3. Tell them that they will be making pollution collectors that they will hang in various places for a week. At the end of the week, the collectors will be analyzed for contents.
4. Have the students assemble in groups of four. Inform them that each will make a pollution collector. There are two different kinds that can be made. Let students decide which kind they want to make.
   - Adhesive paper collector—Get clear adhesive paper and cut it into 10-cm squares. Carefully peel off the backing and tape it to a piece of cardboard, keeping the sticky side out.
   - Wax paper collector—Cut wax paper or plastic wrap into 10 cm-squares. Use tape to secure it to a piece of cardboard. Rub a small amount of petroleum jelly over the surface, using just enough to make it sticky. The paper should look pretty clear when you are done.
5. Have students tape a string loop to the cardboard so the pollution collectors can be hung. Be sure they put their names on the collectors.
6. Let students decide where they are going to hang their collectors. Have them write a description of the location, noting what objects are in the area.
7. Tell students that the collectors will remain in place for three days.

*Final Day*

1. Gather the collectors, marking each one with its location.
2. Ask students to estimate the number of dirt particles on each one.
3. Have them examine the collectors with magnifying lenses and try to identify the different particles.
4. Direct them to complete the student activity sheets.

**Connecting Learning**

1. What kinds of particles were on the collectors?
2. Which collector had the most? ...the least?
3. Which one showed the cleanest air for breathing? Why?
4. Which areas should be avoided?
5. What things contribute to our air pollution?
6. What do you think can be done to make the air cleaner?
7. Were there some particles that would be unsafe to breathe?
8. Were some particles things that shouldn't bother our breathing? What were they?
9. What are you wondering now?

# POLLUTION SOLUTION

## Key Question

What was in the air that you breathe?

## Learning Goal

**Students will:**

observe and identify pollutants in the air.

What do you think is in the air you breathe everyday?

| Setting for collector: | Kind of collector used | Estimate of particles | Kinds of particles |
| --- | --- | --- | --- |
| Area with traffic | | | |
| Pastoral setting | | | |
| Indoors at school | | | |
| Indoors at home | | | |

1. Which collector had the most particles?

   Which had the least?

2. Which area had the cleanest collector?

3. Which areas should be avoided?

POLLUTION SOLUTION
Estimate of particles on collectors
125
120
115
110
105
100
95
90
85
80
75
70
65
60
55
50
45
40
35
30
25
20
15
10
5
0
Traffic Area
Pastoral Setting
Indoors at School
Indoors at Home

CONNECTING LEARNING

## Connecting Learning

1. What kinds of particles were on the collectors?
2. Which collector had the most? ...the least?
3. Which one showed the cleanest air for breathing? Why?
4. Which areas should be avoided?
5. What things contribute to our air pollution?
6. What do you think can be done to make the air cleaner?

 

# POLLUTION SOLUTION

## Connecting Learning

7. Were there some particles that would be unsafe to breathe?

8. Were some particles things that shouldn't bother our breathing? What were they?

9. What are you wondering now?

## Topic
Air quality index

## Key Question
What is the air quality like in the region you researched?

## Learning Goals
Students will:
- learn about the air quality index (AQI) and the pollutants that are measured,
- research a region of the United States to find out their AQI, and
- make inferences as to what might cause the pollutants in that region.

## Guiding Documents
*Project 2061 Benchmarks*
- *Human activities, such as reducing the amount of forest cover, increasing the amount and variety of chemicals released into the atmosphere, and intensive farming, have changed the earth's land, oceans, and atmosphere. Some of these changes have decreased the capacity of the environment to support some life forms.*
- *Locate information in reference books, back issues of newspapers and magazines, compact disks, and computer databases.*

*NRC Standards*
- *Natural environments may contain substances (for example, radon and lead) that are harmful to human beings. Maintaining environmental health involves establishing or monitoring quality standards related to use of soil, water, and air.*
- *Human activities also can induce hazards through resource acquisition, urban growth, land-use decisions, and waste disposal. Such activities can accelerate many natural changes.*

## Science
Earth science
  air pollution
Social sciences
  geography

## Integrated Processes
Observing
Collecting and recording data
Interpreting data
Analyzing
Inferring

## Materials
Internet access
Colored pencils
Student pages

## Background Information
See the *Air Quality Index* fact pages.

## Management
1. Students must have Internet access to complete this activity. They can work in groups of three at a computer.
2. Students will need colored pencils that have green, yellow, orange, red, and purple.

## Procedure
1. Ask the *Key Question* and state the *Learning Goals*.
2. Distribute the *Air Quality Index* fact pages and read through the information as a class, discussing the different pollutants and answering questions as they arise.
3. Tell students that they will be interpreting a map of air quality (specifically, ozone and particles or particulate matter) for various regions of the United States. Give each student a copy of the last student page and colored pencils.
4. Divide the class into groups of three or four students. Tell them that each group will select an area of the United States that they want to research. Then they will look at the AQI for that area for three or four consecutive days, depending on how many are in the group.
5. Provide Internet access and have students go to the Air Now website (http://www.airnow.gov). Once there, direct students to click on the region of their choice. (Discourage them from selecting Hawaii as the archival maps don't show details for Hawaii.) Have them find the current and forecasted data listed for the various cities in that region.

 

Direct them to click on a listed city. This will take them to a page with the map, the AQI, perhaps a forecast discussion. On the left side of that page, toward the bottom, is a link to "Air Quality Maps Archives." Have students click on that and use the back arrow above the calendar to access data for past months. (Summer months will have greater variances in ozone levels.) Encourage students to choose a month, click on the map for a date, and record the data for their region on the last student page. Make sure that each student in the group chooses a consecutive day to record their data.

6. Have members share their maps with each other. Then as a class, have groups share their maps. Compare and contrast the maps and discuss possible reasons for differences.

**Connecting Learning**

1. What is the Air Quality Index? [a measure of pollution in the air]
2. What does the AQI measure? [ground level ozone, particulate matter, carbon monoxide, sulfur dioxide, nitrogen dioxide]
3. What was the air quality like in the region you researched? How do you know?
4. How did the air quality change in the days your group looked at?
5. What factors might have contributed to this level of air quality? [rain, wind, fog, smoke, fires, etc.]
6. How did the air quality in your region compare to the air quality in other regions? What might explain this?
7. What are some things we can do to improve our air quality?
8. What are you wondering now?

**Internet Connections**

*Air Now*

http://www.airnow.gov

The website is used to gather and analyze air quality indices in this activity (see *Procedure 5*). There are interesting links on the left side of the home page under "Learning Center."

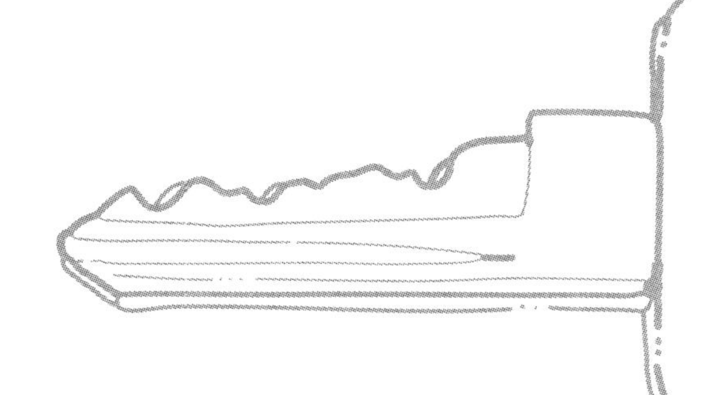

## Key Question

What is the air quality like in the region you researched?

## Learning Goals

**Students will:**

- learn about the air quality index (AQI) and the pollutants that are measured,
- research a region of the United States to find out their AQI, and
- make inferences as to what might cause the pollutants in that region.

The Air Quality Index (AQI) is one of the ways that our government gives us information about our air quality. The index tells us daily how clean or polluted our air is. The AQI is used to help us make decisions about how we can protect our health when the air quality is bad.

Each day a value is given for five major pollutants. The AQI is divided into six categories that give a range of values and the levels of health concern for those values. The AQI is color-coded for ease of use.

| Air Quality Index (AQI) Values | Levels of Health Concern | Colors |
|---|---|---|
| 0 to 50 | Good | Green |
| 51 to 100 | Moderate | Yellow |
| 101 to 150 | Unhealthy for sensitive groups | Orange |
| 151 to 200 | Unhealthy | Red |
| 201 to 300 | Very unhealthy | Purple |
| 301 to 500 | Hazardous | Maroon |

The five major POLLUTANTS that are measured are:

- ground-level ozone,
- particulate matter,
- carbon monoxide,
- sulfur dioxide, and
- nitrogen dioxide.

**Ozone**

Ozone is a gas made up of molecules containing three atoms of oxygen. Ozone can be beneficial or it can be harmful. It depends on where it is found. In the upper atmosphere (the stratosphere), ozone forms a protective layer. This layer helps to protect us from harmful ultraviolet rays from the sun. Without it, we would suffer severe sunburns and have more skin cancer and cataracts. Increased ultraviolet rays also reduce certain crop yields and may harm marine phytoplankton.

 

You have probably heard that the ozone layer is not as thick as it once was. You will even hear reports of large holes in the ozone layer in areas over the Earth's poles. Scientists believe that some chemicals we use are harming this ozone layer. The use of some of these chemicals—chlorofluorocarbons (CFCs)—has been restricted. CFCs were used as refrigerants in air conditioners and refrigerators. They were also used in aerosol cans. Some of these have been replaced with other chemicals and methods in order to protect our atmosphere. Life on Earth needs that protective layer of ozone in the upper atmosphere.

In the lower atmosphere (the troposphere), ozone is a harmful air pollutant. Ground-level ozone is often called smog. It is formed when the emissions from cars, power lawn mowers, leaf blowers, and other gasoline-powered things chemically react with sunlight. Emissions from industries, chemical plants, and power plants can also combine with sunlight to form ozone near the surface of the Earth. It is harmful for us to breathe and can damage plants in areas where the ozone level is high. To help lower the surface ozone level, we should try to limit our use of gasoline-powered things. When we conserve the energy we use in our homes and schools, we are helping to reduce the amount of ozone produced.

**Particulate Matter**

Particulate matter is small particles of liquids or solids that are in the air. Dust, smoke, pollen, molds, and chemicals are examples of particulate matter. Particulate matter can come from our cars and trucks. It can come from fires in fireplaces, forest fires, and agricultural burning. It can come from the dust and smoke from volcanic eruptions. On windy days, particulate matter can come from the dust that is being blown and from the pollen from trees and other plants.

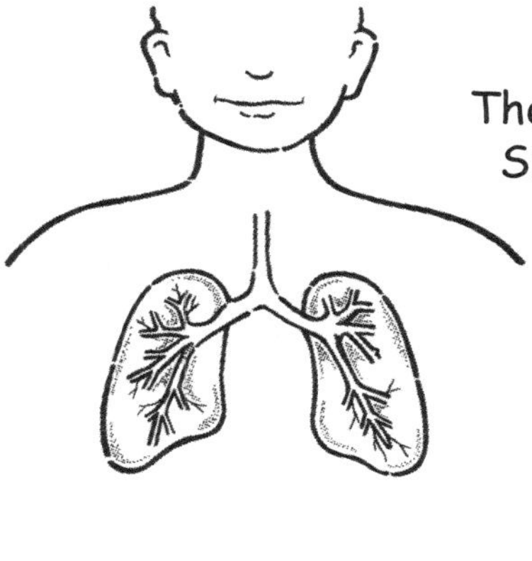

The size of the particulate matter is linked to how harmful it can be. Smaller particles tend to be more harmful than larger particles. Smaller particles can get into our lungs and cause breathing problems. They might even get into our bloodstream and affect our hearts. People with lung diseases, including asthma, are at risk when there are high levels of particulate matter in the air.

 

We can help reduce some of the particulate matter that is in our air. We can reduce our driving. We can also not burn wood in our fireplaces when the air quality is poor. Reducing the amounts of agricultural materials that are burned is another method. However, humans have no control over forest fires caused by lightning or the eruptions of volcanoes. The lesson is that where we can be responsible, we should be.

**Carbon Monoxide**

Carbon monoxide is a poisonous gas. It comes mainly from the exhaust of our gasoline-powered vehicles, lawn mowers, leaf blowers, and chain saws. Other sources are from the burning of fuels and from wildfires. People and animals can die from breathing too much carbon monoxide. Our blood transports oxygen to all parts of our bodies. When we breathe carbon monoxide, it takes the place of oxygen in our blood. We can't live without oxygen. We should be very careful not to be in a closed garage with a vehicle that is running. We should also be very careful that our fireplaces and wood stoves are well vented to the outside.

**Sulfur Dioxide**

Sulfur dioxide is a gas that is produced when coal and oil are burned. Power plants and industry are the major producers of this pollutant. Our nasal passages remove most of the sulfur dioxide we breathe in. However, if we are active enough that we breathe through our mouths, sulfur dioxide can get into our lungs and cause shortness of breath. On days when the sulfur dioxide levels are high, we should reduce our activity levels so that we don't need to breathe through our mouths.

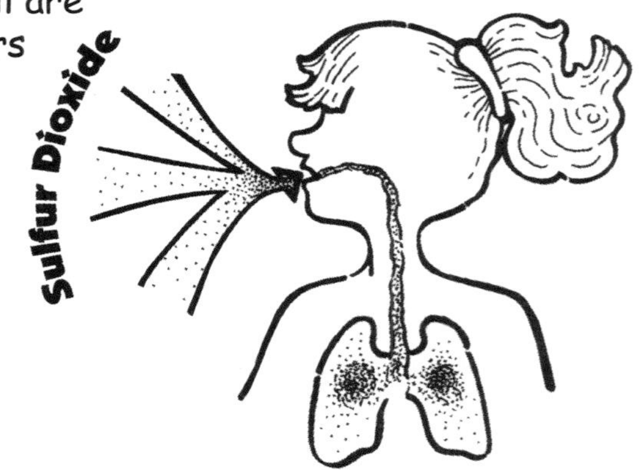


**Nitrogen Dioxide**

Nitrogen dioxide is formed when fuels are burned at high temperatures. Again, motor vehicles, electric utilities, and industries are the major contributors to this type of pollution. Breathing nitrogen dioxide can irritate the lungs and lead to respiratory infections. Conserving our energy use can reduce levels of nitrogen dioxide.

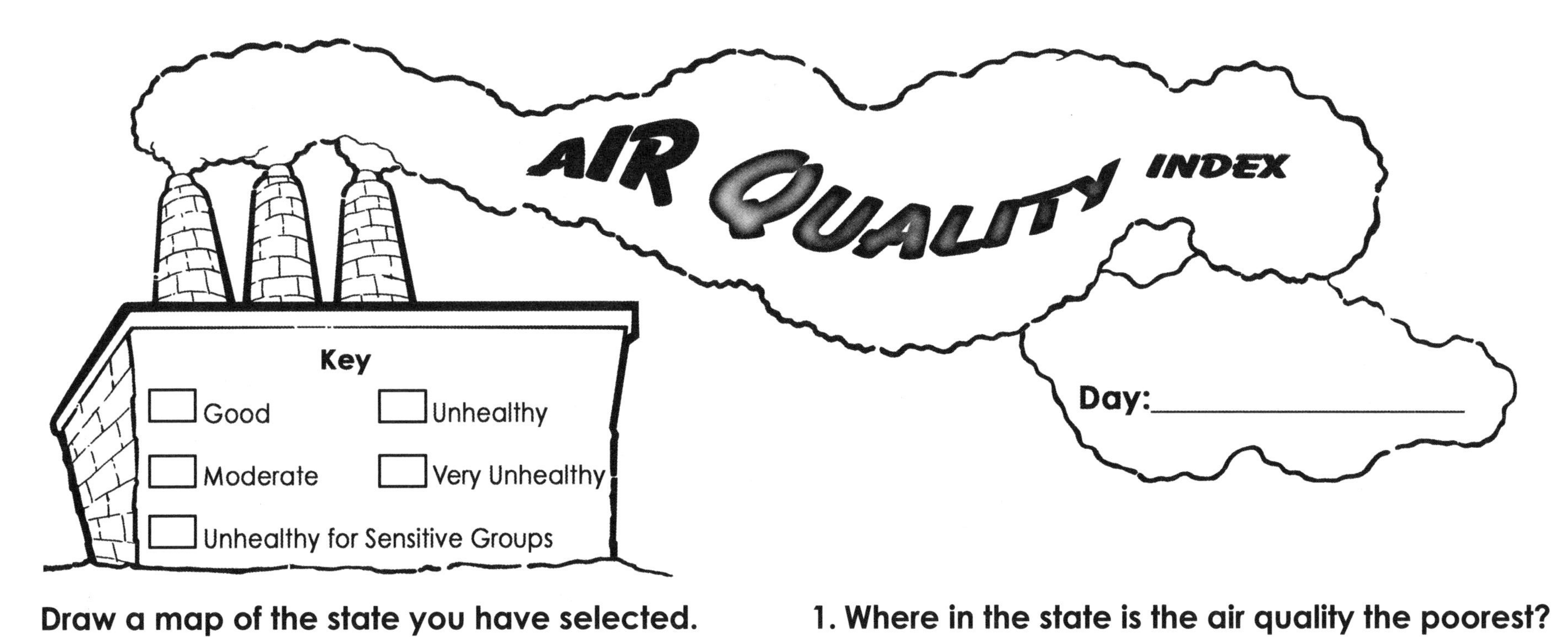


Draw a map of the state you have selected.

1. Where in the state is the air quality the poorest?

2. Where is it the best?

3. Why do you think these conditions exist?

# Connecting Learning

1. What is the Air Quality Index?
2. What does the AQI measure?
3. What was the air quality like in the region you researched? How do you know?
4. How did the air quality change in the days your group looked at?
5. What factors might have contributed to this level of air quality?

## Connecting Learning

6. How did the air quality in your region compare to the air quality in other regions? What might explain this?

7. What are some things we can do to improve our air quality?

8. What are you wondering now?

 

# A-Salting the Environment

**Topic**
Impact of human activities on the environment

**Key Questions**
1. What can humans do to lessen the danger of icy roads?
2. What impact does salting icy roads have on the environment?

**Learning Goals**
Students will:
- investigate a way that is used to keep highways free of ice,
- look at its effects on the environment, and
- discuss how human practices impact the environment.

**Guiding Documents**
*Project 2061 Benchmark*
- *Human activities, such as reducing the amount of forest cover, increasing the amount and variety of chemicals released into the atmosphere, and intensive farming, have changed the earth's land, oceans, and atmosphere. Some of these changes have decreased the capacity of the environment to support some life forms.*

*NRC Standard*
- *Human activities also can induce hazards through resource acquisition, urban growth, land-use decisions, and waste disposal. Such activities can accelerate many natural changes.*

**Math**
Measurement
  volume

**Science**
Physical science
  friction
  chemistry
Life science
  organisms
  environment

**Integrated Processes**
Observing
Comparing and contrasting
Collecting and recording data
Interpreting data
Applying

**Materials**
*Part One*
*Per group:*
  1 or 2 shallow plastic lids covered with ice (see *Management 3)*
  rubber eraser
  salt shaker
  sand, optional

*Part Two*
*Per group:*
  teaspoon
  2 graduated strips
  2 plastic cups, 9 oz
  tape
  salt
  2 small potted salt-sensitive plants (see *Management 6)*
  sticky notes

**Background Information**
Icy winter roads make driving hazardous in many parts of the country. To make these roads safer to drive on, many locations spread salt, sand, or both on the ice. Salt melts the ice and lowers the freezing point of the resulting water significantly, allowing it to run off the road. Sand provides the friction tires need to "grip" icy road surfaces.

These common safety practices, however, come at a cost to the environment. Saltwater runoff can kill roadside plants that are not salt-tolerant. In addition, the runoff can get into nearby bodies of water, raising their salinity levels. This can impact the flora and fauna living there. Although sand is not toxic to plants or animals, it becomes a source of airborne particulate matter that can cause respiratory problems in addition to silting up nearby waterways. It is also a source of damage to the paint and windshields of vehicles.

 

**Management**

1. Both parts of this activity require advanced preparation: making the "ice patches" for *Part One* and growing or purchasing the plants for *Part Two*.
2. *Part One* takes a single class period while *Part Two* extends over several weeks.
3. To make ice patches, fill shallow plastic lids from cottage cheese or similar containers with water (30-50 mL), place them on a cookie sheet, and put them in a freezer. (Note: *Part One* must be started immediately after removing the ice patches from the freezer.)
4. This activity is designed to show the impact salt has on the environment. However, many areas of the country spread sand on icy roads instead of, or in addition to, salt. If desired, students can be given two ice-filled lids instead of one. They will put salt on one and sand on the other to see how both make the ice less slippery. If this option is chosen, twice as many ice patches are needed.
5. *Part Two* examines the effects of salt water on plants over an extended period of time and will require short, daily observations over a period of two weeks.
6. One way to get the plants is to have students grow their own. Radishes are salt-sensitive and germinate and grow quickly, making them ideal for this activity. Other options are beans (most varieties are salt-sensitive) or rye grass (which grows quickly, but is more salt tolerant). If students don't grow their own plants, you can purchase plants at a local nursery. Ask the nursery employees to help you choose inexpensive, salt-sensitive plants. Options include vincas, begonias, ivy, and jasmine. Each group will need two plants, that should be as similar as possible.
7. The plants for *Part Two* should need to be watered the day the experiment starts. These plants should be watered (with fresh and salt water, respectively) as needed, for the duration of the experiment.

**Procedure**

*Part One*

1. Inquire if students have been in a car when the roads were icy. Ask any who have to share their experiences. If no one has, explain that icy roads make driving very dangerous because car tires lose traction and skid easily.
2. Distribute the student page and ask: *What can humans do to lessen the danger of icy roads?*
3. Have students share their ideas on how icy roads are made safer and record these ideas on the student page.
4. Explain that students are going to experience first hand how some areas treat icy roads by spreading salt on these roads.
5. Distribute an ice-filled lid to each group and explain that these represent ice patches on a road. Have students take turns holding the lid stationary on a flat surface and pushing a rubber eraser across the ice. Students should notice that the rubber doesn't grip the ice, but instead slides easily across it. After students have experienced the eraser sliding across the ice, explain that this models how a car tire can skid on an icy road.
6. Distribute the salt shakers. Have students spread a layer of salt on the surface of the ice and carefully observe what happens.
7. After the surface of the ice has begun melting, have students once more push the eraser across the surface. They should notice that it doesn't slide as easily and better grips the surface. This demonstrates how salting icy roads improves traction.
8. If doing the sand option, distribute a second lid and have students repeat the above process, spreading sand, instead of salt, on the ice.
9. Have students discuss the results of the activity, complete the student page, and clean up their areas.

*Part Two*

1. Ask the question: *What impact does salting icy roads have on the environment?*
2. Explain that in this part of the activity, students will set up an ongoing experiment to determine the effect saltwater has on plants. This experiment will have a control plant that is watered with fresh water whenever the soil becomes dry and an experimental plant, which will be watered with salt water.
3. Distribute the plants and materials to each group.
4. Have students use sticky notes or masking tape to label one plant "salt water" and the other "fresh water."
5. Have students carefully observe both plants and record the date and these initial observations on the data sheet. They will leave the *changes* section blank for this initial entry.
6. Distribute two plastic cups and two graduated strips to each group. Direct them to tape the strips to the cups.
7. Have students thoroughly mix a teaspoon of salt with 50 mL of water and pour it on the plant with the salt water label.
8. Have students water the other plant with 50 mL of fresh water.
9. Choose an area of the room like a windowsill or countertop where groups can place their plants for the duration of the experiment. Have them place a sticky note with their names on it next to their group's plants.

 

10. Daily, for the next two weeks (excluding weekends) have students observe their plants, fill in their data sheets, and water plants as necessary using fresh water for the control plant and salt water (1 teaspoon of salt per 50 mL of water) for the experimental plant.
11. At the end of two weeks, have students write their conclusions about the effect salt water has had on their plants. Discuss the connections this experiment has to the practice of spreading salt on icy roads.

**Connecting Learning**

*Part One*

1. Why are icy roads dangerous? [Tires easily lose traction on the ice and the resulting skidding can cause accidents and injuries.]
2. What can humans do to lessen the danger of icy roads? [Various answers: spread salt or sand on the roads, put snow tires or chains on cars, drive carefully, etc.]
3. What did you feel when you pushed the eraser across the ice? [It slid easily across the surface.]
4. Why did the eraser slide? [There is little friction between the ice and the eraser.]
5. How does this sliding relate to cars traveling on icy roads? [The rubber eraser sliding on the icy lid is like a car tire sliding on an icy road.]
6. What did you observe when you spread salt on the ice? [The ice immediately began to melt.]
7. What did you feel when you pushed the eraser across the ice again? [The eraser didn't slide as easily as before. There was more friction.]
8. Why do they spread salt on the roads in some parts of the country? [The salt melts the ice making the roads safer to drive on.]

*Part Two*

1. What things can you observe about your plants and record in your observation log? [Various answers: whether the plants look healthy, the moisture content of the soil, the color of the leaves, the color of the soil, whether the plant is drooping, whether the plant appears to be growing, etc.]
2. What did you observe about your two plants before beginning the experiment? [Both plants looked similar and needed to be watered.]
3. What things did you observe about the plant you watered with salt water? [Various answers: The salt formed a white deposit on the soil surface. The leaves turned yellowish. The plant looked sick. The plant began to droop. The plant withered. Etc.]
4. What did you observe about the plant you watered with fresh water? [Various answers: The plant looked healthy. The plant stayed green. The plant grew. Etc.]
5. What can you say about the effect of salt water on your plant? [Salt water is not good for the plant.]
6. What might this activity say about the impact salting icy roads has on the environment? [The salt water that runs off roads into the surrounding area may damage some of the plants living there.]
7. What are some positive aspects of salting icy roads? [Various answers: Human lives may be saved since there are few accidents. Money is saved since there are fewer accidents. Salt-tolerant plants thrive on shoulders of the roads. Etc.]
8. What are some negative aspects of salting icy roads? [Various answers: Salt-sensitive plants are killed or damaged. Saltwater runoff may get in local waterways impacting the plants and animals living there. Salt water can cause cars to rust. Etc.]
9. Explain how salting the roads is an example of how a solution to one problem may create other problems. [The salting makes roads safer to drive on, but has a negative impact on parts of the environment.]
10. What are you wondering now?

**Extensions**

1. Have students repeat the experiment with different concentrations of salt water.
2. Have students use the Internet to research salt-tolerant plants.

# A-Salting the Environment

## Key Question

1. What can humans do to lessen the danger of icy roads?
2. What impact does salting icy roads have on the environment?

## Learning Goals

**Students will:**

- investigate a way that is used to keep highways free of ice,
- look at its effects on the environment, and
- discuss how human practices impact the environment.

# A-Salting the Environment

## Graduated Scales
### for 9-oz cups

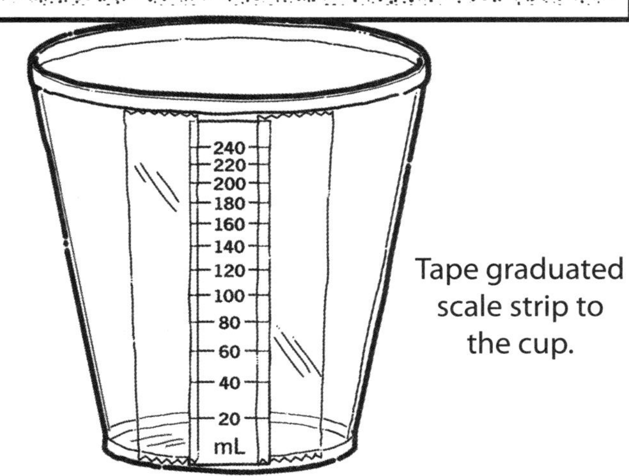


Tape graduated scale strip to the cup.

| 240 | 240 | 240 | 240 | 240 | 240 | 240 | 240 | 240 | 240 |
|---|---|---|---|---|---|---|---|---|---|
| 220 | 220 | 220 | 220 | 220 | 220 | 220 | 220 | 220 | 220 |
| 200 | 200 | 200 | 200 | 200 | 200 | 200 | 200 | 200 | 200 |
| 180 | 180 | 180 | 180 | 180 | 180 | 180 | 180 | 180 | 180 |
| 160 | 160 | 160 | 160 | 160 | 160 | 160 | 160 | 160 | 160 |
| 140 | 140 | 140 | 140 | 140 | 140 | 140 | 140 | 140 | 140 |
| 120 | 120 | 120 | 120 | 120 | 120 | 120 | 120 | 120 | 120 |
| 100 | 100 | 100 | 100 | 100 | 100 | 100 | 100 | 100 | 100 |
| 80 | 80 | 80 | 80 | 80 | 80 | 80 | 80 | 80 | 80 |
| 60 | 60 | 60 | 60 | 60 | 60 | 60 | 60 | 60 | 60 |
| 40 | 40 | 40 | 40 | 40 | 40 | 40 | 40 | 40 | 40 |
| 20 | 20 | 20 | 20 | 20 | 20 | 20 | 20 | 20 | 20 |
| mL | mL | mL | mL | mL | mL | mL | mL | mL | mL |

| 240 | 240 | 240 | 240 | 240 | 240 | 240 | 240 | 240 | 240 |
|---|---|---|---|---|---|---|---|---|---|
| 220 | 220 | 220 | 220 | 220 | 220 | 220 | 220 | 220 | 220 |
| 200 | 200 | 200 | 200 | 200 | 200 | 200 | 200 | 200 | 200 |
| 180 | 180 | 180 | 180 | 180 | 180 | 180 | 180 | 180 | 180 |
| 160 | 160 | 160 | 160 | 160 | 160 | 160 | 160 | 160 | 160 |
| 140 | 140 | 140 | 140 | 140 | 140 | 140 | 140 | 140 | 140 |
| 120 | 120 | 120 | 120 | 120 | 120 | 120 | 120 | 120 | 120 |
| 100 | 100 | 100 | 100 | 100 | 100 | 100 | 100 | 100 | 100 |
| 80 | 80 | 80 | 80 | 80 | 80 | 80 | 80 | 80 | 80 |
| 60 | 60 | 60 | 60 | 60 | 60 | 60 | 60 | 60 | 60 |
| 40 | 40 | 40 | 40 | 40 | 40 | 40 | 40 | 40 | 40 |
| 20 | 20 | 20 | 20 | 20 | 20 | 20 | 20 | 20 | 20 |
| mL | mL | mL | mL | mL | mL | mL | mL | mL | mL |

 

# A-Salting the Environment

## Part One

**What can humans do to lessen the danger of icy roads?**
Discuss your ideas with your group and record them here.

Place the "ice patch" your teacher gives you on a flat surface. While holding the edge of the plastic lid, push down on an eraser as you move it across the surface of the ice. What do you notice?

How does this part of the activity relate to cars traveling on icy roads?

Spread a layer of salt on the ice. Watch what happens and record your observations below.

After a few minutes, push your eraser across the surface once more. What did you notice this time?

Why do they spread salt on icy roads in some parts of the country?

# A-Salting the Environment

## Part Two

**What impact does salting icy roads have on the environment?**

In Part One you observed how spreading salt on ice caused it to melt. This practice makes icy roads safer, but impacts the areas around the roads when the salty water drains off. In this part, you will do an experiment to observe the effect saltwater has on certain plants.

To set up this scientific experiment, you need two plants: an experimental plant and a control plant. The experimental plant will be watered with salt water while the control plant will be watered with fresh water.

Use sticky tabs to label one plant "salt water" and the other "fresh water."

Carefully observe both plants and record these observations and the date on the data sheet. Make sure to include such things as the color of the plant leaves, the color of the soil, the moisture level of the soil, whether the plants look healthy or sick, etc.

Thoroughly mix one teaspoon of salt with 50 mL of water and pour this salt water on the plant labeled "salt water."

Pour 50 mL of fresh water on the remaining plant.

Place the plants in an area where they can be observed for the next two weeks. Write the names of the people in your group on sticky notes and place these notes next to your plants.

Make daily observations of both plants for a period of at least two weeks. Record all of your observations on the data sheets, making special note of any changes observed.

At the end of two weeks, draw your conclusions about the impact salt water had on your plants. Record your conclusions in your science journal.

 

# A-Salting the Environment

## Observation Sheet

Observation number:
Date:
Description of plant with salt water:

Changes, if any, from the previous observation:

Description of plant with fresh water:

Changes, if any, from the previous observation:

Observation number:
Date:
Description of plant with salt water:

Changes, if any, from the previous observation:

Description of plant with fresh water:

Changes, if any, from the previous observation:

# A-Salting the Environment

## Connecting Learning

*Part One*

1. Why are icy roads dangerous?

2. What can humans do to lessen the danger of icy roads?

3. What did you feel when you pushed the eraser across the ice?

4. Why did the eraser slide?

5. How does this sliding relate to cars traveling on icy roads?

6. What did you observe when you spread salt on the ice?

7. What did you feel when you pushed the eraser across the ice again?

8. Why do they spread salt on the roads in some parts of the country?

 

# A-Salting the Environment

## Connecting Learning

*Part Two*

1. What things can you observe about your plants and record in your observation log?
2. What did you observe about your two plants before beginning the experiment?
3. What things did you observe about the plant you watered with salt water?
4. What did you observe about the plant you watered with fresh water?
5. What can you say about the effect of salt water on your plant?
6. What might this activity say about the impact salting icy roads has on the environment?

CONNECTING LEARNING

## Connecting Learning

7. What are some positive aspects of salting icy roads?

8. What are some negative aspects of salting icy roads?

9. Explain how salting the roads is an example of how a solution to one problem may create other problems.

10. What are you wondering now?

**Topic**
Pollution and its effect on animals

**Key Question**
How can pollution affect the survival of animals?

**Learning Goals**
Students will:
- play a game to simulate the effect of pollution on a population of foxes, and
- relate this simulation to real-world situations.

**Guiding Documents**
*Project 2061 Benchmark*
- *Changes in an organism's habitat are sometimes beneficial to it and sometimes harmful.*

*NRC Standards*
- *Changes in environments can be natural or influenced by humans. Some changes are good, some are bad, and some are neither good nor bad. Pollution is a change in the environment that can influence the health, survival, or activities of organisms, including humans.*
- *Environments are the space, conditions, and factors that affect an individual's and a population's ability to survive and their quality of life.*

**Math**
Data analysis
graphing

**Science**
Environmental science
pollution

**Integrated Processes**
Observing
Collecting and recording data
Interpreting data
Generalizing

**Materials**
*For each student:*
*Pollution's Effect on Wildlife* rubber band book
#19 rubber band

*For each group:*
paper bag
*Fox Pieces* (see *Management 2* and *3)*
*Pollution Pieces* (see *Management 2* and *3)*
game mat (see *Management 4)*
scissors

**Background Information**

In this activity, students play a simple game that simulates the effect of pollution on a population of foxes.The activity begins with students placing 10 healthy fox pieces and two pollution pieces in a paper bag. The bag is shaken to mix the pieces. Without looking into the bag, students draw two pieces and place them on game mats that picture the local environment. If both pieces are foxes, they are placed in the *Forest.* If one piece is a fox and the other a unit of pollution, they are placed in the *Polluted Stream.* If both pieces are pollution units, they go in the *Factory Pollution* section of the mat. If the last draw consists of a single fox or pollutant piece, it is placed in the *Single Draw Pieces* section of the second game mat. After all pieces have been drawn and classified, the appropriate replacements are made (see the rules page) and the new set of pieces is placed in the bag. This process of drawing, classifying, replacing, and placing the pieces back into the bag is called a *generation.* With only two units of pollution, the population of foxes increases.

Students play the game a second time with the initial pollution level doubled to four units. The higher pollution level causes much greater damage to the fox population. In this game, the fox population decreases and may even die out.

The game is based on simple probability. Two pollutant pieces mixed with 10 fox pieces means that picking two fox pieces is more likely to happen than picking a fox and a pollution piece or two pollution pieces. At this low pollution level, the population is likely to increase. When the number of pollution pieces is doubled, it becomes more likely that the population of foxes decreases or even dies out.

 

**Management**

1. Students can work in pairs or groups of four.
2. Print the *Fox Pieces* and *Pollution Pieces* on different colors of card stock. Each group will need at least one *Fox Pieces* page and at least half of a *Pollution Pieces* page. Have extra *Fox Pieces* and *Pollution Pieces* available in case groups need more counters.
3. Area Tiles, Teddy Bear Counters, or other counters can be used in place of the *Fox Pieces* and *Pollution Pieces.* Two colors are needed and both should be the same size and shape so there is no distinction other than color when they are drawn from the paper bag. Area Tiles (item number 4810) and Teddy Bear Counters (item number 1924) are available from AIMS.
4. Make one game mat for each group by taping the two included pages together to make one large page.

**Procedure**

1. Ask the *Key Question* and invite students to think of any examples they have heard of—in current events or from many years ago—where pollution has harmed animal populations. [2010 Deepwater Horizon oil spill, 1989 Exxon Valdez oil spill, plastic trash harming ocean creatures, etc.]
2. Distribute the rubber band book and a #19 rubber band to each student. Read through the book as a class and discuss the additional effects of pollution described.
3. Tell students that they are now going to simulate a scenario similar to the one that happened with DDT. Hand out scissors, *Fox Pieces*, and *Pollution Pieces* to each group. Have the students cut out the individual pieces.
4. Give each group a copy of the rules page and the game mat. Use the rules page to explain and model the process students are to use to complete one generation of the game.
5. Distribute the first student page to each student. Help students complete drawing and classifying the pieces for the first generation. Work through the replacements with the class as a whole to make sure the rules are understood. Repeat the draw, classify, and replace process for as many generations as needed to clarify the rules.
6. Have students complete all five generations.
7. Distribute the second student page to each student. Have students complete all five generations of the game with increased pollution.
8. If you want your students to graph the data from the two tables, distribute the third student page to every student.
9. Discuss the results and how the change to the environment—pollution—affected the population of foxes. Relate the experience to real-world situations students have read and heard about.

**Connecting Learning**

1. With two units of pollution, what happened to the foxes living in the forest? [The number of foxes (population) got larger.]
2. What happened to the foxes living in the forest when the pollution was doubled? [The number of foxes got smaller.]
3. What are some things that pollute where the animals live? [Chemicals from farms and factories that get into the water, soil, and air; smog from automobiles; etc.]
4. How does this game show you what might happen in the real world? [When pollution levels get high, plants and animals suffer harm and maybe death.]
5. What are some sources of pollution in our area?
6. How can we lower the levels of pollution?
7. What are you wondering now?

## Key Question

How can pollution affect the survival of animals?

## Learning Goals

**Students will:**

- play a game to simulate the effect of pollution on a population of foxes, and
- relate this simulation to real-world situations.

 

# Pollution's Effects on Wildlife

According to the dictionary, *pollution* is "the presence in or introduction into the environment of a substance or thing that has harmful or poisonous effects." Something that causes pollution is called a *pollutant.*

1

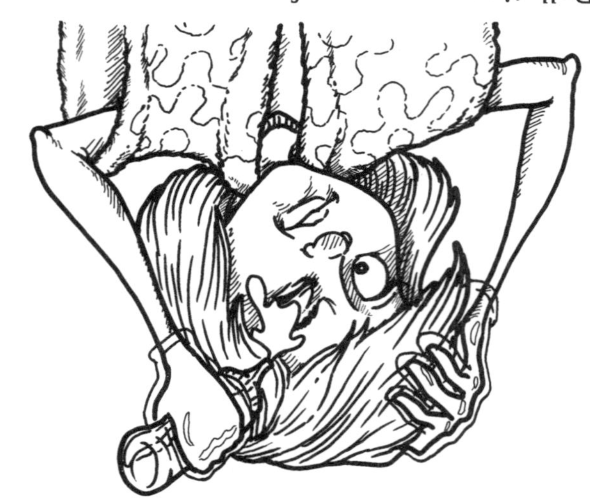

Pollution can come from many sources. Some comes from chemicals we spray on crops and lawns to kill weeds or bugs. Some comes from factories, trains, and automobiles. Even common household products like kitchen cleaners and hair dye contain chemicals that can be pollutants.

2

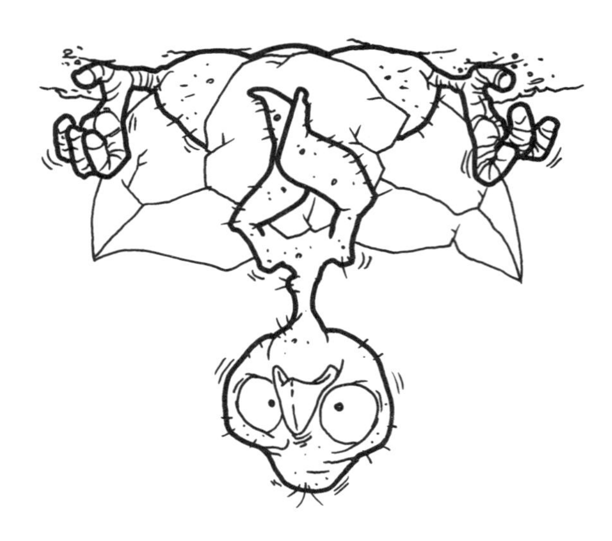

DDT was banned in the United States in 1972. Since then, most bird populations have come back. Still, over 35 years later, California Condors continue to have problems with thin shells.

7

Sadly, we did not learn our lesson with DDT. There continue to be many pollutants in the environment that have harmful effects on humans and animals. We need to carefully study the effects of chemicals and balance benefits and risks. To learn more, check out the World Wildlife Fund: wwf.panda.org.

8

When these chemicals and other pollutants get into soil, air, and water, they have harmful effects.

3

In the first part of the 20th century, a pesticide called DDT was used to kill mosquitoes and other insects. (The goal was to control diseases they carried.) It was often sprayed from the air.

4

Soon, scientists began to notice harmful effects on the environment. Many other animals besides the insects were being affected. DDT was especially harmful to birds like the bald eagle, brown pelican, peregrine falcon, and osprey.

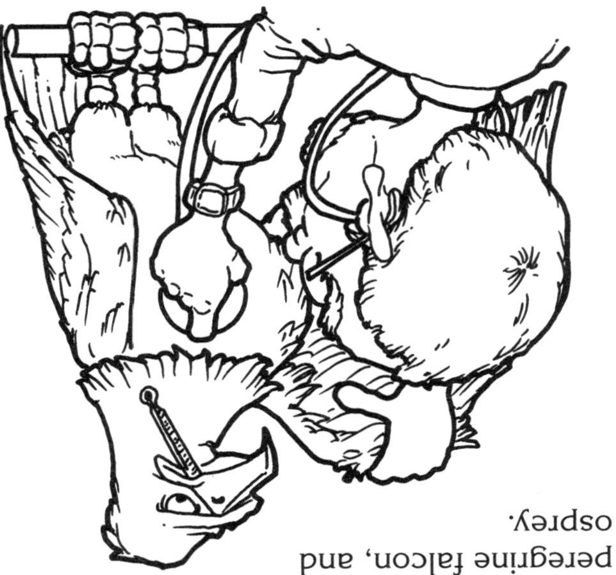

5

The chemical made the shells of the birds' eggs thin. The eggs would crack before the baby birds were ready to hatch. The bald eagle almost went extinct.

6

## Fox Pieces

Copy this page onto card stock. Each group needs one page.

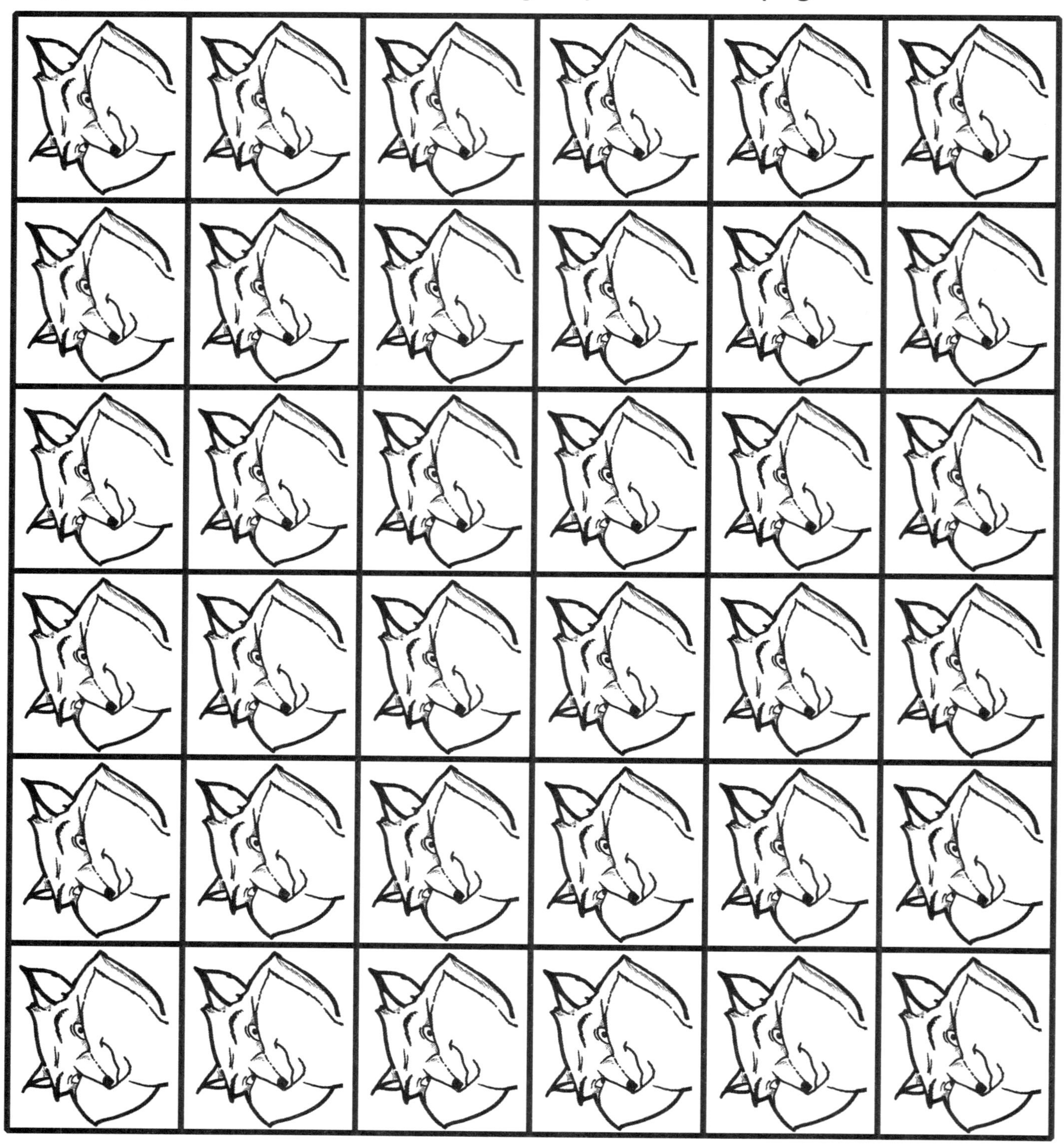

## Pollution Pieces

Copy this page onto a different color of card than the *Fox Pieces*. Each group needs half of a page.

Forest
Polluted Stream
Foxes in the Forest

Factory
Pollution
Graveyard
R.I.P.
Single Draw Pieces

 

# Foxes in the Forest

## Draw

Put the number of foxes and pollutants listed at the top of the data table in the bag. Shake the bag and, without looking, draw two pieces from the bag.

## Classify

IF…

the pair are both foxes, stack the pair in the ***Forest.***

one is a fox and the other is a pollutant, stack the pair in the ***Polluted Stream.***

only one counter is left at the last draw, put that counter in the ***Single Draw Pieces.***

the pair are both pollutants, stack them in the ***Factory Pollution.***

## Replace

Move each fox in the ***Polluted Stream*** to the ***Fox Graveyard.*** The fox died because of the pollution.

For each pair of foxes in the ***Forest,*** add one fox.

For each pair of pollutants in the ***Factory Pollution,*** add two more pollutants.

## Repeat

Take all of the foxes and pollutants from the forest, stream, and factory and put them back into the bag. Leave all dead foxes in the graveyard and single draw pieces on the game mat. Shake the bag to mix them before playing the next generation.

# Foxes in the Forest

Put 10 fox pieces and two pollution pieces in the bag and play the game.

| Generation | Starting # of foxes | Starting # of pollutants | # of foxes that died | # of healthy foxes |
|---|---|---|---|---|
| 1 | 10 | 2 | | |
| 2 | | | | |
| 3 | | | | |
| 4 | | | | |
| 5 | | | | |

What happened to the foxes living in the forest?

 

# Foxes in the Forest

Put 10 fox pieces and four pollution pieces in the bag and play the game.

| Generation | Starting # of foxes | Starting # of pollutants | # of foxes that died | # of healthy foxes |
|---|---|---|---|---|
| 1 | 10 | 4 | | |
| 2 | | | | |
| 3 | | | | |
| 4 | | | | |
| 5 | | | | |

What happened to the foxes living in the forest after the pollution doubled?

# Foxes in the Forest

Make a bar graph to show the number of healthy foxes at the end of each generation. Use one color for the first scenario and a different color for the second scenario. Be sure to fill in the key to show the colors.

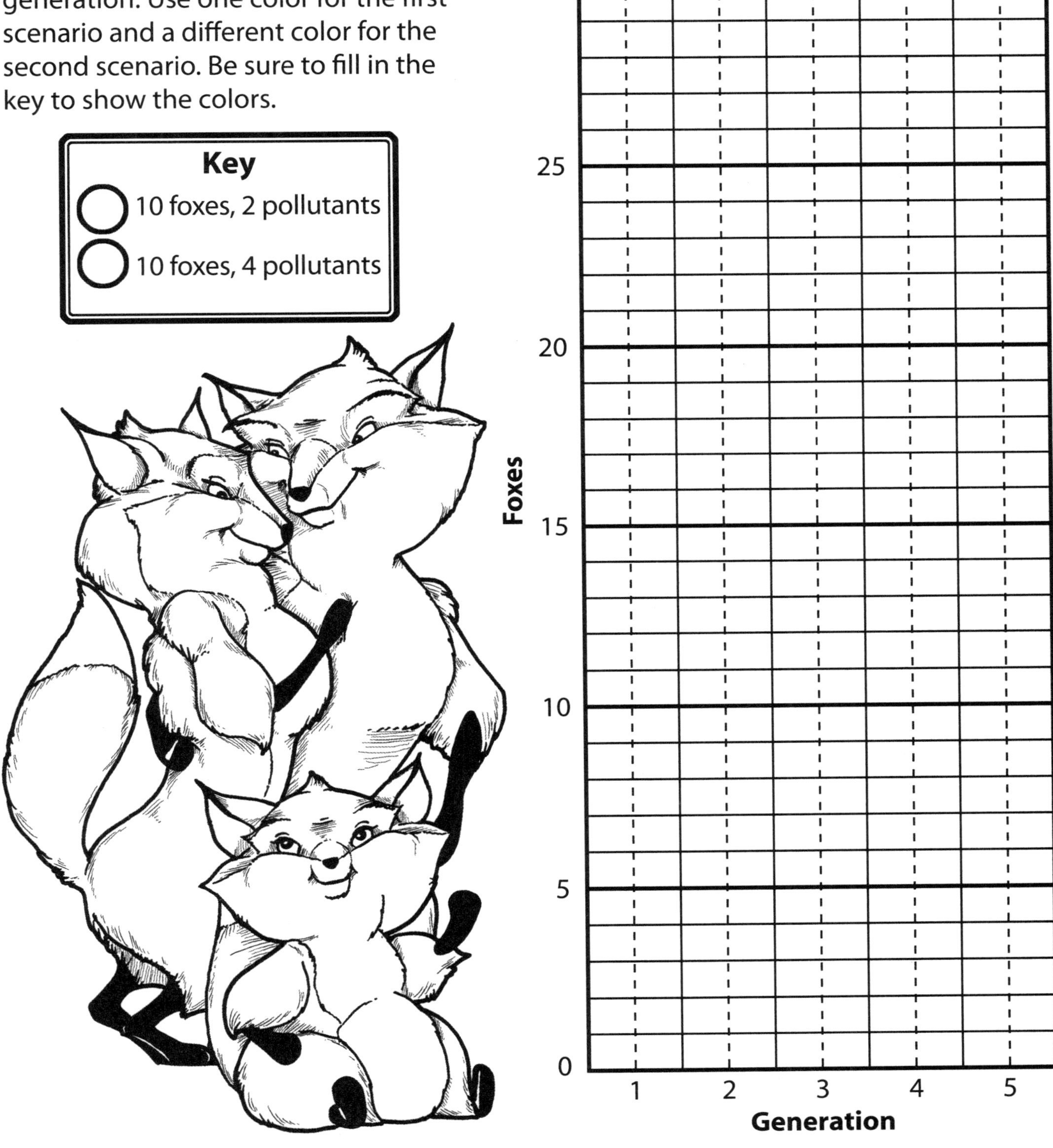

CONNECTING LEARNING

## Connecting Learning

1. With two units of pollution, what happened to the foxes living in the forest?
2. What happened to the foxes living in the forest when the pollution was doubled?
3. What are things that pollute where the animals live?
4. How does this game show you what might happen in the real world?
5. What are some sources of pollution in our area?

## Connecting Learning

6. How can we lower the levels of pollution?

7. What are you wondering now?

# Water Rights and Responsibilities

**Topic**
River systems (watersheds)

**Key Question**
How does land use affect river systems?

**Learning Goals**
Students will:
- identify land uses that affect river systems, and
- determine if the impact is positive or negative.

**Guiding Documents**
*Project 2061 Benchmark*
- *In something that consists of many parts, the parts usually influence one another.*

*NRC Standards*
- *Risk analysis considers the type of hazard and estimates the number of people that might be exposed and the number likely to suffer consequences. The results are used to determine the options for reducing or eliminating risks.*
- *Students should understand the risks associated with natural hazards (fires, floods, tornadoes, hurricanes, earthquakes, and volcanic eruptions), with chemical hazards (pollutants in air, water, soil, and food), with biological hazards (pollen, viruses, bacterial, and parasites), social hazards (occupational safety and transportation), and with personal hazards (smoking, dieting, and drinking).*
- *Individuals can use a systematic approach to thinking critically about risks and benefits. Examples include applying probability estimates to risks and comparing them to estimated personal and social benefits.*
- *Important personal and social decisions are made based on perceptions of benefits and risks.*

**Science**
Earth science
river systems (watersheds)

**Integrated Processes**
Observing
Analyzing
Drawing conclusions

**Materials**
*For the class:*
bulletin board paper (see *Management 2)*
large raindrop

*For each student group:*
*Land Use Description Cards* (see *Management 3)*
*Land Use Location Clue Cards (*see *Management 3)*
pink and green sticky notes, 3-inch squares
colored pencils
white construction paper, 12 x 18-inch
white copy paper

*For each student*
rubber band book
rubber band, #19

**Background Information**
See rubber band book.

**Management**
1. Divide the class into eight groups.
2. Prepare blue bulletin board paper with the following illustration and labels, and position it on the wall where it is easily seen.

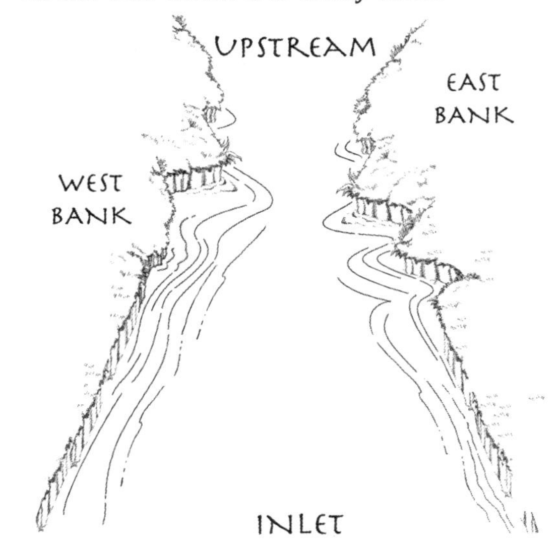


3. Prepare a set *Land Use Description Cards,* a large raindrop on the white construction paper, and a set *Land Use Location Clue Cards.* Each group will need one location card and one description card.

 

4. These are the correct placements for the *Land Use Locations*.

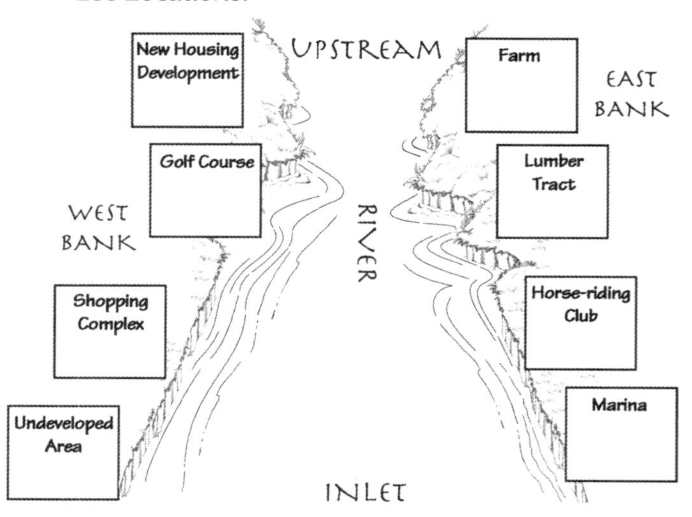


**Procedure**

1. Ask the *Key Question* and state the *Learning Goals.*
2. Distribute the rubber band book *Water Rights and Responsibilities*. Direct the students to read and discuss the information in it.
3. Draw students' attention to the blue bulletin board paper and tell the students that it represents a model river system.
4. Discuss different ways that land is used. [shopping malls, farming, housing, etc.] Distribute the *Land Use Description* cards, colored pencils, and copy paper. Direct each group to illustrate the land use described on their card. Tell them to be sure to label their pictures with the site's use (e.g., Lumber Tract, Farm, Shopping Complex, etc.).
5. Discuss with the students how most land uses have both positive and negative impacts on the environment and us. Challenge the students to create a list of the positive and negative impacts caused by their group's land use.
6. Distribute the pink and green sticky notes. Have students record each negative impact caused by their land use on a pink sticky note, and each positive impact caused by their land use on a green sticky note. Each impact statement should be listed on a separate note. Have students attach the sticky notes to their completed drawings.
7. Give each group one *Land Use Location Clue Card*. Explain that each group will read their clue aloud and that the class will determine the correct placement for each illustration based on the clues.
8. When the class has decided where each illustration should be located, check for correct placement. Tape or glue the illustrations to the paper river picture.
9. Show the students the large raindrop and explain that it will represent the water that moves through a river system. As the raindrop passes each site, read the sticky notes. As you attach the sticky notes to the drop of water, discuss whether the positive or negative impacts of each land use outweigh each other.
10. When the raindrop has completed its journey down the river, discuss what possible changes could be made to lessen negative impacts. For example, by planting the corn on the farm in rows parallel to the river, there would be less run-off into the river.

**Connecting Learning**

1. What did you learn about how land uses around a river impact the river?
2. How did you decide if the negative impact is less important than the positive impact?
3. What are some ways we can reduce negative impacts on the river system?
4. If you live on a river, why should you care about what is happening upstream or downstream from you?
5. What are some reasons we are willing to accept the consequences of negative impacts?
6. What are you wondering now?

 

# Water Rights and Responsibilities

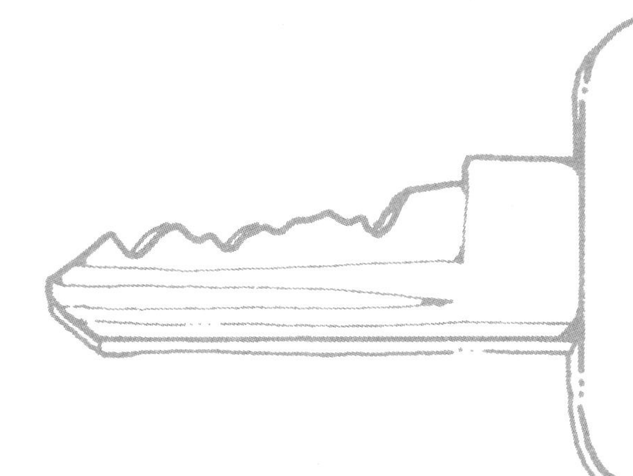

**Key Question**

How does land use affect river systems?

## Learning Goals

**Students will:**

- identify land uses that affect river systems, and
- determine if the impact is positive or negative.

 

# Water Rights and Responsibilities

## Land Use Location Clue Cards

### Clue Card 1

The marina is located on the East Bank closest to the Inlet.

### Clue Card 2

The lumber tract is on the opposite bank of the undeveloped area but not directly across from it.

### Clue Card 3

The golf course is on the same side as the undeveloped area.

### Clue Card 4

The marina is not next to the farm, but is on the same side of the river.

### Clue Card 5

The shopping complex is between the undeveloped area and the golf course.

### Clue Card 6

The new housing development is furthest upsteam on the West Bank of the river.

### Clue Card 7

The lumber tract is between the horse-riding club and the farm.

### Clue Card 8

The new housing development is beside the golf course.

# Water Rights and Responsibilities

## Land Use Description Cards

### Land Use Site—Farm

A farmer owns this section of the river. Her family has raised corn on this land for more than 100 years. Corn takes a large amount of nutrients out of the soil. A large quantity of fertilizer is added to the soil every growing season. Fertilizer is sprayed onto the plants themselves when they are first beginning to grow. The corn is planted in rows perpendicular to the river. Many of the rows slope down close to the river's edge.

### Land Use Site—Lumber Tract

A company owns this section of the river on which pine trees grow. These trees were planted 50 years ago and the company wants to harvest them and sell them to a lumber company. The company usually sends in three or four large bulldozers to plow through the underbrush. This allows quick and easy access to the trees. The bulldozers do not have any devices to help trap pollutants. It is company policy to always replant areas that are harvested through clear cutting.

### Land Use Site—Marina

There is a large marina on this section of the river. It has four large concrete and asphalt docks and ramps that slope directly down to the river. This allows boats easy access to the water. The marina sells oil and gasoline to boaters from the dock closest to the river.

### Land Use Site—New Housing Development

This section of the river is scheduled for development. There will be 25 new homes. The developer has paved three streets as well as installed sidewalks. The drainage system is designed to flow into the local storm drains. The plans call for cutting down the largest trees beside the water so that the new homeowners will have a better view of the river.

 

# Water Rights and Responsibilities

## Land Use Description Cards

### Land Use Site—Shopping Complex

This section of land on the river has a large shopping complex on it. The developer has large grass areas and trees to create a park-like atmosphere. The parking lots are located away from the edge of the river. Many of the shops are located very close to the river. A large concrete walkway borders the river's edge so that people can walk and enjoy the river.

### Land Use Site—Golf Course

This section of the river has an 18-hole golf course. The developer kept most of the large trees that lined the river's edge. The greens and fairways are kept green by frequent applications of fertilizers and herbicides. Many of the fairways slope down to the river's edge.

### Land Use Site—Horse-riding Club

There is a horse-riding club along this section of the river. One large grassed pasture borders the river. There are many trails winding through undeveloped land with native plants and trees. The barn that houses the horses is beside the river. A large manure pile from cleaning out the barn is kept beside the river.

### Land Use Site—Undeveloped Area

This section of the river is an undeveloped tract that contains native plants and trees.

 

7

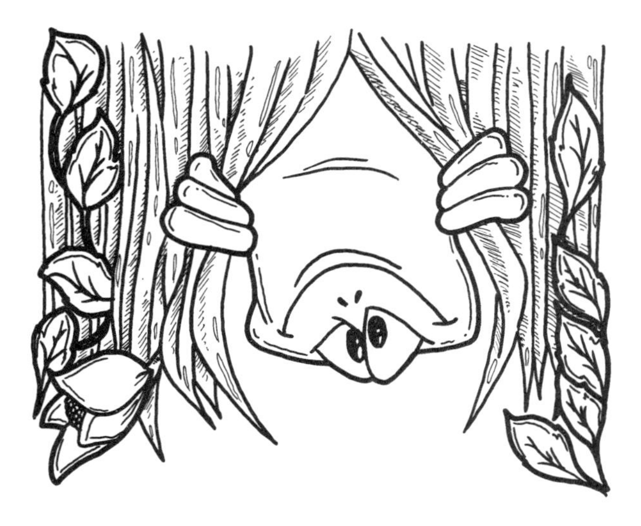

Maintaining a plant zone around bodies of water can reduce runoff. Plants serve as a natural filter. They also slow water runoff. Trees provide shade. This helps keep the water from reaching unhealthy temperatures.

Farmers who use no-till farming and contour plowing help to reduce erosion and runoff. Developers who maintain plant life along waterways help reduce the potential runoff of pollutants and sediment. Runoff from paved areas can be channeled so that it does not directly enter the river system. Working with nature benefits us all.

8

2

The health of a river basin system is affected by the things entering its waters. The condition of the land next to rivers, streams, and creeks is important to its health. However, it is not only these lands that have an effect.

1

 

Activities far from rivers or other bodies of water can impact the system's health. When it rains, the water that does not soak into the ground is carried into bodies of water. This runoff often adds sediments, extra nutrients, and pollutants to the water.

3

Sediment is dirt that is carried along by water. When it enters a river or stream, it blocks sunlight. This prevents underwater plants from getting the light they need. With less light, the plants produce less oxygen. Low oxygen levels harm the water animals that need oxygen to survive.

4

Excess nutrients can come from fertilizers and animal waste. These nutrients are a food source for algae. If too many nutrients are present, an algal bloom can develop. This has the same effect as sediment in the water. It blocks sunlight and reduces oxygen.

5

Pollutants can also enter river systems as runoff. Motor oil, road salt, and other chemicals are toxic. They can kill fish and other animal life in rivers, lakes, and streams.

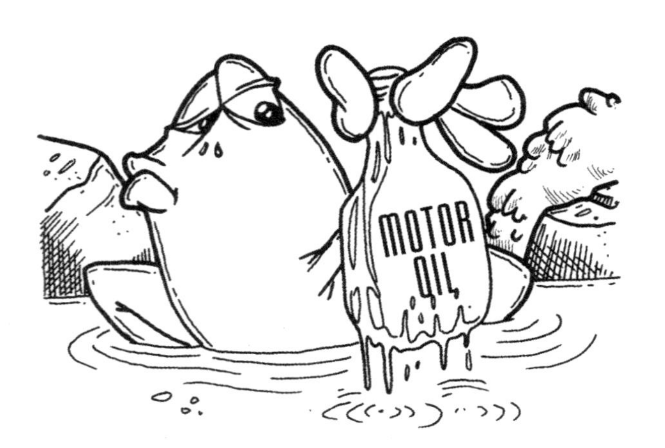


6

 

# Water Rights and Responsibilities

CONNECTING LEARNING

## Connecting Learning

1. What did you learn about how land uses around a river impact the river?

2. How did you decide if the negative impact is less important than the positive impact?

3. What are some ways we can reduce negative impacts on the river system?

4. If you live on a river, why should you care about what is happening upstream or downstream from you?

5. What are some reasons we are willing to accept the consequences of negative impacts?

6. What are you wondering now?

 

# Earth Rocks!

## Materials List

### Equipment

* Rock kits (#4108R)
* Mineral kits (#4108M)
* Hand lenses (#1977)
* Ring magnets (#1971)
* Eyedroppers (#1946)
* Balances (#1917)
* Masses (#1923)
Safety goggles
Digital camera

*Available from AIMS

### Consumables and Non-consumables

Pencils
Crayons
Colored pencils
Glue sticks
White glue
Scissors
Erasers
Tape
Sticky notes, 3-inch squares—pink, green
White copy paper
White contruction paper, 12" x 18"
Chalk
Hole punch
Bulletin board paper
Rubber bands, #19
Empty one-liter bottles with lids
Empty two-liter bottles
Soil
Potting soil
Sand
Pebbles
Peat moss
Radish seeds
Funnels
Clay
Baby food jars
Chenille stems
Index cards, 3" x 5"
Epsom salts
Cookie sheet
Aluminum foil
Cardboard
Crispy rice cereal
Miniature marshmallows
Raisins
Toffee chips
Candy-coated chocolate candies
Steel nail file
Sandpaper
10-penny nails
Unglazed white tiles
Vinegar
Tablespoon measure
Teaspoon measure
Toothpicks
Wax paper
Styrofoam cups, 5 oz
Plastic cups, 16-20 oz
Plastic cups, 10 oz
Plastic cups, 9 oz
Plastic cups, 3-3.5 oz
Paper cups, 3 oz
Plastic spoons
Plastic knives
Plastic bowls
Paper plates
Zipper-type plastic bags, gallon size
Zipper-type plastic bags, quart size
Nut brittle
Nuts
Flaked coconut
Sunflower seeds
Brown sugar
Coffee filter, basket type
Glitter
Cardboard box with lid
Plastic trash bag, kitchen size
Plastic spray bottle
Paper towels
Food coloring
Plaster of Paris
Disposable lasagna pans
Disposable pie plate
Duct tape
Drinking straws
Styrofoam peanuts
Newspapers
Shoebox lid
Hand-held hair dryer
Potted plant
Sugar cubes
White bread
Whole-wheat bread
Dark rye bread
Jam or jelly
Peanut butter
Paper lunch bag
Clear adhesive paper
Petroleum jelly
Salt in shaker

 

# The AIMS Program

AIMS is the acronym for "**A**ctivities **I**ntegrating **M**athematics and **S**cience." Such integration enriches learning and makes it meaningful and holistic. AIMS began as a project of Fresno Pacific University to integrate the study of mathematics and science in grades K-9, but has since expanded to include language arts, social studies, and other disciplines.

AIMS is a continuing program of the non-profit AIMS Education Foundation. It had its inception in a National Science Foundation funded program whose purpose was to explore the effectiveness of integrating mathematics and science. The project directors, in cooperation with 80 elementary classroom teachers, devoted two years to a thorough field-testing of the results and implications of integration.

The approach met with such positive results that the decision was made to launch a program to create instructional materials incorporating this concept. Despite the fact that thoughtful educators have long recommended an integrative approach, very little appropriate material was available in 1981 when the project began. A series of writing projects ensued, and today the AIMS Education Foundation is committed to continuing the creation of new integrated activities on a permanent basis.

The AIMS program is funded through the sale of books, products, and professional-development workshops, and through proceeds from the Foundation's endowment. All net income from programs and products flows into a trust fund administered by the AIMS Education Foundation. Use of these funds is restricted to support of research, development, and publication of new materials. Writers donate all their rights to the Foundation to support its ongoing program. No royalties are paid to the writers.

The rationale for integration lies in the fact that science, mathematics, language arts, social studies, etc., are integrally interwoven in the real world, from which it follows that they should be similarly treated in the classroom where students are being prepared to live in that world. Teachers who use the AIMS program give enthusiastic endorsement to the effectiveness of this approach.

Science encompasses the art of questioning, investigating, hypothesizing, discovering, and communicating. Mathematics is a language that provides clarity, objectivity, and understanding. The language arts provide us with powerful tools of communication. Many of the major contemporary societal issues stem from advancements in science and must be studied in the context of the social sciences. Therefore, it is timely that all of us take seriously a more holistic method of educating our students. This goal motivates all who are associated with the AIMS Program. We invite you to join us in this effort.

Meaningful integration of knowledge is a major recommendation coming from the nation's professional science and mathematics associations. The American Association for the Advancement of Science in *Science for All Americans* strongly recommends the integration of mathematics, science, and technology. The National Council of Teachers of Mathematics places strong emphasis on applications of mathematics found in science investigations. AIMS is fully aligned with these recommendations.

Extensive field testing of AIMS investigations confirms these beneficial results:

1. Mathematics becomes more meaningful, hence more useful, when it is applied to situations that interest students.
2. The extent to which science is studied and understood is increased when mathematics and science are integrated.
3. There is improved quality of learning and retention, supporting the thesis that learning which is meaningful and relevant is more effective.
4. Motivation and involvement are increased dramatically as students investigate real-world situations and participate actively in the process.

We invite you to become part of this classroom teacher movement by using an integrated approach to learning and sharing any suggestions you may have. The AIMS Program welcomes you!

# Get the Most From Your Hands-on Teaching

When you host an AIMS workshop for elementary and middle school educators, you will know your teachers are receiving effective, usable training they can apply in their classrooms immediately.

**AIMS Workshops are Designed for Teachers**

- Hands-on activities
- Correlated to your state standards
- Address key topic areas, including math content, science content, and process skills
- Provide practice of activity-based teaching
- Address classroom management issues and higher-order thinking skills
- Include $50 of materials for each participant
- Offer optional college (graduate-level) credits

**AIMS Workshops Fit District/Administrative Needs**

- Flexible scheduling and grade-span options
- Customized workshops meet specific schedule, topic, state standards, and grade-span needs
- Sustained staff development can be scheduled throughout the school year
- Eligible for funding under the Title I and Title II sections of No Child Left Behind
- Affordable professional development—consecutive-day workshops offer considerable savings

**Call us to explore an AIMS workshop**
**1.888.733.2467**

**Online and Correspondence Courses**

AIMS offers online and correspondence courses on many of our books through a partnership with Fresno Pacific University.

- Study at your own pace and schedule
- Earn graduate-level college credits

## See all that AIMS has to offer—visit us online

 **http://www.aimsedu.org** 

Check out our website where you can:

- preview and purchase AIMS books and individual activities;
- learn about State-Specific Science and Essential Math;
- explore professional development workshops and online learning opportunities;
- buy manipulatives and other classroom resources; and
- download free resources including articles, puzzles, and sample AIMS activities.

**find us on facebook**

Become a fan of AIMS!

- Be the first to hear of new products and programs.
- Get links to videos on using specific AIMS lessons.
- Join the conversation—share how you and your students are using AIMS.

While visiting the AIMS website, sign up for our FREE *AIMS for You* e-mail newsletter to get free activities, puzzles, and subscriber-only specials delivered to your inbox monthly.

# AIMS Program Publications

Actions With Fractions, 4-9
The Amazing Circle, 4-9
Awesome Addition and Super Subtraction, 2-3
Bats Incredible! 2-4
Brick Layers II, 4-9
The Budding Botanist, 3-6
Chemistry Matters, 5-7
Counting on Coins, K-2
Cycles of Knowing and Growing, 1-3
Crazy About Cotton, 3-7
Critters, 2-5
Earth Book, 6-9
Earth Explorations, 2-3
Earth, Moon, Sun, 3-5
Earth Rocks! 4-5
Electrical Connections, 4-9
Energy Explorations: Sound, Light, and Heat, 3-5
Exploring Environments, K-6
Fabulous Fractions, 3-6
Fall Into Math and Science*, K-1
Field Detectives, 3-6
Floaters and Sinkers, 5-9
From Head to Toe, 5-9
Getting Into Geometry, K-1
Glide Into Winter With Math and Science*, K-1
Gravity Rules! 5-12
Hardhatting in a Geo-World, 3-5
Historical Connections in Mathematics, Vol. I, 5-9
Historical Connections in Mathematics, Vol. II, 5-9
Historical Connections in Mathematics, Vol. III, 5-9
It's About Time, K-2
It Must Be A Bird, Pre-K-2
Jaw Breakers and Heart Thumpers, 3-5
Looking at Geometry, 6-9
Looking at Lines, 6-9
Machine Shop, 5-9
Magnificent Microworld Adventures, 6-9
Marvelous Multiplication and Dazzling Division, 4-5
Math + Science, A Solution, 5-9
Mathematicians are People, Too
Mathematicians are People, Too, Vol. II
Mostly Magnets, 3-6
Movie Math Mania, 6-9
Multiplication the Algebra Way, 6-8
Out of This World, 4-8
Paper Square Geometry:
The Mathematics of Origami, 5-12
Popping With Power, 3-5
Positive vs. Negative, 6-9
Primarily Bears*, K-6
Primarily Magnets, K-2
Primarily Physics: Investigations in Sound, Light, and Heat Energy, K-2
Primarily Plants, K-3
Primarily Weather, K-3
Probing Space, 3-5
Problem Solving: Just for the Fun of It! 4-9
Problem Solving: Just for the Fun of It! Book Two, 4-9
Proportional Reasoning, 6-9
Puzzle Play, 4-8
Ray's Reflections, 4-8
Sensational Springtime, K-2
Sense-able Science, K-1
Shapes, Solids, and More: Concepts in Geometry, 2-3
Simply Machines, 3-5
The Sky's the Limit, 5-9
Soap Films and Bubbles, 4-9
Solve It! K-1: Problem-Solving Strategies, K-1
Solve It! 2nd: Problem-Solving Strategies, 2
Solve It! 3rd: Problem-Solving Strategies, 3
Solve It! 4th: Problem-Solving Strategies, 4
Solve It! 5th: Problem-Solving Strategies, 5
Solving Equations: A Conceptual Approach, 6-9
Spatial Visualization, 4-9
Spills and Ripples, 5-12
Spring Into Math and Science*, K-1
Statistics and Probability, 6-9
Through the Eyes of the Explorers, 5-9
Under Construction, K-2
Water, Precious Water, 4-6
Weather Sense: Temperature, Air Pressure, and Wind, 4-5
Weather Sense: Moisture, 4-5
What on Earth? K-1
What's Next, Volume 1, 4-12
What's Next, Volume 2, 4-12
What's Next, Volume 3, 4-12
Winter Wonders, K-2

Essential Math
Area Formulas for Parallelograms, Triangles, and Trapezoids, 6-8
Circumference and Area of Circles, 5-7
Effects of Changing Lengths, 6-8
Measurement of Prisms, Pyramids, Cylinders, and Cones, 6-8
Measurement of Rectangular Solids, 5-7
Perimeter and Area of Rectangles, 4-6
The Pythagorean Relationship, 6-8

* Spanish supplements are available for these books. They are only available as downloads from the AIMS website. The supplements contain only the student pages in Spanish; you will need the English version of the book for the teacher's text.

For further information, contact:
AIMS Education Foundation • P.O. Box 8120 • Fresno, California 93747-8120
www.aimsedu.org • 559.255.6396 (fax) • 888.733.2467 (toll free)

# Duplication Rights

No part of any AIMS books, magazines, activities, or content—digital or otherwise—may be reproduced or transmitted in any form or by any means except as noted below.

## Standard Duplication Rights

- A person or school purchasing AIMS activities (in books, magazines, or in digital form) is hereby granted permission to make up to 200 copies of any portion of those activities, provided these copies will be used for educational purposes and only at one school site.
- For a workshop or conference session, presenters may make one copy of any portion of a purchased activity for each participant, with a limit of five activities or up to one-third of a book, whichever is less.
- All copies must bear the AIMS Education Foundation copyright information.
- Modifications to AIMS pages (e.g., separating page elements for use on an interactive white board) are permitted only within the classroom or school for which they were purchased, or by presenters at conferences or workshops. Interactive white board files may not be uploaded to any third-party website or otherwise distributed. AIMS artwork and content may not be used on non-AIMS materials.

Standard duplication rights apply to activities received at workshops, free sample activities provided by AIMS, and activities received by conference participants.

## Unlimited Duplication Rights

Unlimited duplication rights may be purchased in cases where AIMS users wish to:

- make more than 200 copies of a book/magazine/activity,
- use a book/magazine/activity at more than one school site, or
- make an activity available on the Internet (see below).

These rights permit unlimited duplication of purchased books, magazines, and/or activities (including revisions) for use at a given school site.

Activities received at workshops are eligible for upgrade from standard to unlimited duplication rights.

Free sample activities and activities received as a conference participant are not eligible for upgrade from standard to unlimited duplication rights.

State-Specific Science modules are licensed to one classroom/one teacher and are therefore not eligible for upgrade from standard to unlimited duplication rights.

## Upgrade Fees

The fees for upgrading from standard to unlimited duplication rights are as follows.
For individual activities, the cost is $5 per activity per school site.
For Literature Links bundles, the cost is $12 per bundle per school site.
For books, the cost is based on the price of the book (see table).

| Book Price | Upgrade Fee |
|---|---|
| $9.95 | $15.00/site |
| $18.95 | $24.00/site |
| $21.95 | $27.00/site |
| $24.95 | $30.00/site |
| $29.95 | $35.00/site |
| $34.95 | $40.00/site |
| $49.95 | $55.00/site |

The cost of upgrading is shown in the following examples:
For five activities at six schools:
5 activities x $5 x 6 schools = $150

For two books (at $21.95) at 10 schools:
2 books x $27 x 10 schools = $540

For three books (at $24.95) and four activities at eight schools:
(3 books x $30 x 8 schools) + (4 activites x $5 x 8 schools) = $720 + $160 = $880

## Purchasing Unlimited Duplication Rights

To purchase unlimited duplication rights, please provide us the following:

1. The name of the individual responsible for coordinating the purchase of duplication rights.
2. The title of each book, activity, and/or magazine issue to be covered.
3. The number of school sites and name and address of each site for which rights are being purchased.
4. Payment (check, purchase order, credit card).

Requested duplication rights are automatically authorized with payment. The individual responsible for coordinating the purchase of duplication rights will be sent a certificate verifying the purchase.

## Internet Use

AIMS materials may be made available on the Internet if all of the following stipulations are met:

1. The materials to be put online are purchased as PDF files from AIMS (i.e., no scanned copies).
2. Unlimited duplication rights are purchased for all materials to be put online for each school at which they will be used. (See above.)
3. The materials are made available via a secure, password-protected system that can only be accessed by employees at schools for which duplication rights have been purchased.

AIMS materials may not be made available on any publicly accessible Internet site.